T5-BBY-347

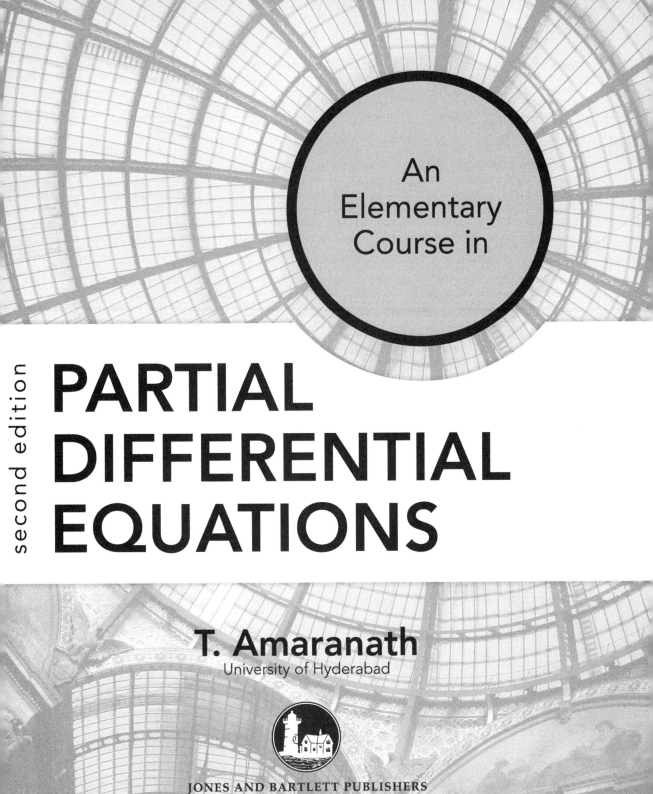

second edition

An
Elementary
Course in

# PARTIAL
# DIFFERENTIAL
# EQUATIONS

## T. Amaranath
University of Hyderabad

**JONES AND BARTLETT PUBLISHERS**
*Sudbury, Massachusetts*
BOSTON    TORONTO    LONDON    SINGAPORE

*World Headquarters*
Jones and Bartlett Publishers
40 Tall Pine Drive
Sudbury, MA 01776
978-443-5000
info@jbpub.com
www.jbpub.com

Jones and Bartlett Publishers
Canada
6339 Ormindale Way
Mississauga, Ontario L5V 1J2
Canada

Jones and Bartlett Publishers
International
Barb House, Barb Mews
London W6 7PA
United Kingdom

Jones and Bartlett's books and products are available through most bookstores and online booksellers. To contact Jones and Bartlett Publishers directly, call 800-832-0034, fax 978-443-8000, or visit our website www.jbpub.com.

Substantial discounts on bulk quantities of Jones and Bartlett's publications are available to corporations, professional associations, and other qualified organizations. For details and specific discount information, contact the special sales department at Jones and Bartlett via the above contact information or send an email to specialsales@jbpub.com.

*An Elementary Course in Partial Differential Equations, Second Edition*, originally published by © 2003 Narosa Publishing House, New Delhi—110 002. All Rights Reserved.

Copyright © 2009 by Jones and Bartlett Publishers, LLC

All rights reserved. No part of the material protected by this copyright may be reproduced or utilized in any form, electronic or mechanical, including photocopying, recording, or by any information storage and retrieval system, without written permission from the copyright owner.

Not for sale outside of the United States of America, Canada, and Puerto Rico. Export or sale of this book outside of the United States of America, Canada, and Puerto Rico is illegal.

**Production Credits**
Acquisitions Editor: Timothy Anderson
Editorial Assistant: Melissa Potter
Production Director: Amy Rose
Senior Marketing Manager: Andrea DeFronzo
V.P., Manufacturing and Inventory Control: Therese Connell
Composition: Northeast Compositors, Inc.
Cover Design: Kristin E. Ohlin
Cover Image: © Mschalke/Dreamstime.com
Printing and Binding: Malloy, Inc.
Cover Printing: Malloy, Inc.

**Library of Congress Cataloging-in-Publication Data**
Amaranath, T.
    An elementary course in partial differential equations / T. Amaranath. — 2nd ed.
        p. cm.
    Includes bibliographical references and index.
    ISBN-13: 978-0-7637-6244-5 (hardcover)
    ISBN-10: 0-7637-6244-X (ibid.)
    1. Differential equations, Partial. I. Title.

QA377.A5663 2008
515'.353—dc22

200804

6048

Printed in the United States of America
12 11 10 09 08    10 9 8 7 6 5 4 3 2 1

*Dedicated to the memory of my parents*

# Preface

There are a large number of books written on partial differential equations. Yet there is no single book available that can be used as a textbook for a one-semester course. Having faced this problem every time I taught a course on partial differential equations, I felt compelled to prepare lecture notes that could be circulated among my students. Their enthusiastic and overwhelming response toward them has been the main reason for bringing forth the Indian edition of the book a decade ago.

The book consists of two chapters, each having many sections. The first chapter deals with first order partial differential equations while the second one is devoted to the study of second order partial differential equations.

While preparing this book I tried to make the presentation of the various topics comprehensive and rigorous, yet simple. However, students must be careful whenever inverse or implicit function theorems are used in the book, as the results obtained are valid only locally even though it is not mentioned explicitly every time.

In almost all the topics covered herein, various examples have been worked out and many exercises have been provided at the end of each topic along with the final answers in many places.

Converting lecture notes into a book involves a lot of effort and I would like to thank my colleague Dr. B. Sri Padmavati, without whose sincere help this would not have been possible. I also thank Professor C. Musili for having taken a personal interest in getting the Indian edition of this book published. In addition, I would like to take this opportunity to express my sincere thanks to my teachers Professor S. D. Nigam and Professor K. V. Naik to whom I am highly indebted. Special thanks are due as well to Professor C. N. Kaul who spared his valuable time to go through the final draft of this book.

Thanks are due to my wife, Beena, and my sons, Prashanth and Karteek, for their steady encouragement, support, and care.

Finally, I thank Narosa Publishing House and Jones and Bartlett Publishers for publishing this edition.

T. AMARANATH

# Contents

# Chapter 1

# First Order Partial Differential Equations

## 1.1 Curves and Surfaces

This section briefly describes two geometrical objects, curves and surfaces. They both play a very important role in the study of partial differential equations.

**Curves in space**: A curve may be specified by means of parametric equations. Suppose $f_1, f_2$, and $f_3$ are continuous functions of a continuous variable $t$ that varies in an interval $I \subseteq \mathbb{R}$, then the three equations $x = f_1(t), y = f_2(t), z = f_3(t)$ represent the parametric equations of a curve in three-dimensional space.

A standard parameter is the length of the curve measured from some fixed point on the curve. In such cases, at times, the symbol $s$ is used instead of $t$.

**Note**: The condition that the parameter $t$ is the length of the curve is $f_1'^2 + f_2'^2 + f_3'^2 = 1$, where prime denotes differentiation.

**Example 1.1.1**: The simplest example of a curve in space is a straight line with direction cosines $(l, m, n)$ passing through a point $(x_0, y_0, z_0)$ with parametric equations

$$x = x_0 + ls, y = y_0 + ms, z = z_0 + ns, \quad s \in \mathbb{R}. \qquad \square$$

**Example 1.1.2**: A right circular helix is a space curve lying on a circular cylinder and is given by the following parametric equations

$$x = a \cos wt, y = a \sin wt, z = kt, \quad t \in \mathbb{R},$$

where $a, w$, and $k$ are constants. □

**Surfaces**: A point $(x, y, z)$ in space is said to lie on a surface if the coordinates $x, y$, and $z$ satisfy

$$F(x, y, z) = 0, \tag{1.1.1}$$

where $F$ is a continuously differentiable function defined on a domain in $\mathbb{R}^3$, and $\frac{\partial F}{\partial x}, \frac{\partial F}{\partial y}$, and $\frac{\partial F}{\partial z}$ do not vanish simultaneously. Therefore a surface is the locus of a point moving in space with "two degrees of freedom". Consider a set of relations of the form

$$x = F_1(u, v), y = F_2(u, v), z = F_3(u, v). \tag{1.1.2}$$

Then for each pair of values of $u$ and $v$ there correspond three numbers $x, y$, and $z$ and hence a point $(x, y, z)$ in space. It is, however, not true that every point in space corresponds to a pair $u$ and $v$. If the Jacobian $\dfrac{\partial(F_1, F_2)}{\partial(u, v)} = \dfrac{\partial F_1}{\partial u}\dfrac{\partial F_2}{\partial v} - \dfrac{\partial F_1}{\partial v}\dfrac{\partial F_2}{\partial u} \neq 0,$ then the first two equations of (1.1.2) can be solved to express $u$ and $v$ as functions of $x$ and $y$ locally, say, $u = \lambda(x, y)$ and $v = \mu(x, y)$ by the inverse function theorem. Then $u$ and $v$ are determined once $x$ and $y$ are known and the third equation in (1.1.2) gives a value of $z$ for these values of $x$ and $y$. That is

$$z = F_3(\lambda(x, y), \mu(x, y)),$$

which is thus a functional relation between the coordinates $x$, $y$, and $z$ as in (1.1.1). Thus any point $(x, y, z)$ determined from Equation (1.1.2) always lies on a fixed surface. For this reason, equations of the type (1.1.2) are called parametric equations of the surface. However, parametric equations of a surface are not unique as the following example demonstrates.

**Example 1.1.3**: The parametric equations

$$x = a \sin u \cos v, \ y = a \sin u \sin v, \ z = a \cos u,$$

and $\qquad x = a\dfrac{(1 - v^2)}{(1 + v^2)} \cos u \ , \ y = a\dfrac{(1 - v^2)}{(1 + v^2)} \sin u \ , \ z = \dfrac{2av}{(1 + v^2)} \ ,$

where $a$ is a constant, both represent the surface $x^2 + y^2 + z^2 = a^2$, which is a sphere. □

Let us now go back to the equation of the curve

$$x = f_1(t), \quad y = f_2(t), \quad z = f_3(t).$$

On eliminating $t$ between $f_1$ and $f_2$, we obtain a relation $\phi(x, y) = 0$. Similarly, from $f_1$ and $f_3$ we get $\psi(x, z) = 0$. Hence a curve can be thought of as the points of intersection of two surfaces. If the parameter $s$ in the equation of a curve is the length of the curve measured from some fixed point, then $(\frac{dx}{ds}, \frac{dy}{ds}, \frac{dz}{ds})$ are the direction cosines of the tangent to the curve at that point $s$.

Suppose that the curve $C \colon (x(s), y(s), z(s))$ lies on the surface $S$ whose equation is $F(x, y, z) = 0$. Then

$$F(x(s), y(s), z(s)) \equiv 0 \ \forall \ s.$$

On differentiating with respect to $s$, we get

$$\frac{\partial F}{\partial x}\frac{dx}{ds} + \frac{\partial F}{\partial y}\frac{dy}{ds} + \frac{\partial F}{\partial z}\frac{dz}{ds} = 0.$$

This is the condition that the tangent to the curve $C$ at the point $P(x, y, z)$ is perpendicular to the line with direction ratios $(\frac{\partial F}{\partial x}, \frac{\partial F}{\partial y}, \frac{\partial F}{\partial z})$. It may be noted that the curve $C$ is arbitrary except for the fact that it passes through the point $P$ and lies on the given surface. Therefore we can conclude that the line whose direction ratios are given by $(\frac{\partial F}{\partial x}, \frac{\partial F}{\partial y}, \frac{\partial F}{\partial z})$ is perpendicular to the tangent to every curve lying on $S$ and passing through $P$, and therefore they give the direction ratios of the normal to the surface $S$ at the point $P$.

## 1.2 Genesis of First Order P.D.E.

By a partial differential equation (p.d.e.), we mean an equation of the form

$$f(x, y, t, \cdots, \theta, \theta_x, \theta_y, \cdots, \theta_{xt}, \cdots) = 0, \tag{1.2.1}$$

involving two or more independent variables $x, y, t$, etc., one dependent variable $\theta(x, y, t, \cdots) \in C^n$ in some domain $D$ and its partial derivatives $\theta_x$, $\theta_y$, $\cdots$, $\theta_{xx}$, $\theta_{xt}$, $\cdots$, where $C^n$ denotes a set of functions possessing continuous partial derivatives of order $n$. A p.d.e. is thus a relation between the dependent variable $\theta$ and some of its partial derivatives at every point $(x, y, t, \cdots)$ in $D$.

**Definition 1.2.1**: We define the order of a partial differential equation to be the order of the derivative of the highest order $n$ occurring in the equation.

A p.d.e. is said to be *quasi-linear* if the derivatives of the highest order that occur in the equation are linear. A quasi-linear p.d.e. is said to be *semi-linear* if the coefficients of the highest order derivatives do not contain either the dependent variable or its derivatives. A semi-linear p.d.e. is said to be linear if it is linear in the dependent variable and its derivatives (please refer to pages 7 and 75).

A p.d.e. that is not a quasi-linear p.d.e. is said to be *non-linear*.

We are concerned in this chapter with partial differential equations of first order with one dependent variable $z$ and mostly two independent variables $x$ and $y$. We follow the notation $p = z_x$ and $q = z_y$. We can then write the most general first order p.d.e. in the form

$$f(x, y, z, p, q) = 0.$$

Partial differential equations arise in a large variety of subjects: in geometry, physics, mathematics, etc.

**Example 1.2.1: Surfaces of revolution**

Many surfaces of revolution with $z$-axis as the axis of revolution are of the form

$$z = F(r), \quad r = (x^2 + y^2)^{\frac{1}{2}},$$

where $F$ is an arbitrary continuously differentiable function.

On differentiating $z = F(r)$ with respect to $x$ and $y$ respectively, we obtain

$$p = (\frac{x}{r})F'(r), \quad q = (\frac{y}{r})F'(r).$$

After eliminating the function $F'(r)$, we get

$$yp - xq = 0,$$

which is a p.d.e. of first order satisfied by the surfaces of revolution of the form given previously. □

**Example 1.2.2**: Consider the surfaces of the form $F(u, v) = 0$ where $u = u(x, y, z)$ and $v = v(x, y, z)$ are known functions of $x, y$, and $z$, and $F$ is an arbitrary function of $u$ and $v$ having first order partial derivatives with respect to $u$ and $v$.

On differentiating $F(u, v) = 0$ with respect to $x$ and $y$, treating $z$ as a function of $x$ and $y$, we get respectively

$$\frac{\partial F}{\partial u}(u_x + pu_z) + \frac{\partial F}{\partial v}(v_x + pv_z) = 0,$$

$$\frac{\partial F}{\partial u}(u_y + qu_z) + \frac{\partial F}{\partial v}(v_y + qv_z) = 0.$$

On eliminating $\dfrac{\partial F}{\partial u}$ and $\dfrac{\partial F}{\partial v}$ from these equations, we obtain

$$p\frac{\partial(u,v)}{\partial(y,z)} + q\frac{\partial(u,v)}{\partial(z,x)} = \frac{\partial(u,v)}{\partial(x,y)},$$

which is a p.d.e. of first order satisfied by the surface $F(u,v) = 0$, where

$$\frac{\partial(u,v)}{\partial(x,y)} = u_x v_y - u_y v_x,$$

is the Jacobian. □

**Example 1.2.3: Functional dependence**

Let $v = v(x,y)$ be a known function of $x$ and $y$ and let $u = u(x,y)$ be a function of $v$ alone, i.e., not involving $x$ and $y$ explicitly. That is

$$u = H(v) \,,$$

where $H$ is arbitrary. Let $H \in C^1$.

On differentiating the previous equation with respect to $x$ and $y$ respectively, we get

$$u_x = H'(v)v_x \,, \quad u_y = H'(v)v_y.$$

On eliminating $H'(v)$ from these equations, we get

$$u_x v_y - u_y v_x = \frac{\partial(u,v)}{\partial(x,y)} = 0,$$

which is a first order p.d.e. for $u$. □

**Note**: Hence if there is a functional relation between two functions $u(x,y)$ and $v(x,y)$ not involving $x$ and $y$ explicitly, then

$$\frac{\partial(u,v)}{\partial(x,y)} = 0.$$

The converse of this result is proved in Lemma 1.5.1.

**Example 1.2.4: Euler's equation for a homogeneous function**

A function $f(x,y)$ is said to be a homogeneous function of $x$ and $y$ of degree $n$ if it satisfies

$$f(\lambda x, \lambda y) = \lambda^n f(x,y).$$

Let $z = f(x, y)$ be a homogeneous function of $x$ and $y$ of degree $n$. Then by Euler's theorem, the function $f(x, y)$ satisfies the first order p.d.e.

$$xf_x + yf_y = nf. \qquad \square$$

**Note**: In each of Examples 1.2.1 to 1.2.3, we have produced a p.d.e. by eliminating an arbitrary function.

Now consider a two-parameter family of surfaces

$$z = F(x, y, a, b). \qquad (1.2.2)$$

On differentiating with respect to $x$ and $y$, we obtain respectively

$$p = F_x(x, y, a, b), \qquad (1.2.3)$$

$$q = F_y(x, y, a, b). \qquad (1.2.4)$$

We can solve two of the previous three equations to find $a$ and $b$ in terms of $x, y, p$, and $q$. This would be possible provided that the matrix

$$\begin{pmatrix} F_a & F_{xa} & F_{ya} \\ F_b & F_{xb} & F_{yb} \end{pmatrix} \qquad (1.2.5)$$

is of rank two by the implicit function theorem. Substituting $a$ and $b$ thus obtained in the third equation will usually lead to a relation of the form $f(x, y, z, p, q) = 0$, which is a first order p.d.e.

**Example 1.2.5**: Consider

$$z = x + ax^2y^2 + b. \qquad (1.2.6)$$

On differentiating (1.2.6) with respect to $x$ and $y$, we get respectively

$$p = 1 + 2axy^2, \qquad (1.2.7)$$

$$q = 2ax^2y. \qquad (1.2.8)$$

On eliminating $a$ between (1.2.7) and (1.2.8), we obtain

$$xp - yq - x = 0. \qquad \square$$

**Example 1.2.6**: Consider

$$(x - a)^2 + (y - b)^2 + z^2 = 1. \qquad (1.2.9)$$

On differentiating (1.2.9) with respect to $x$ and $y$, we get respectively

$$zp = -(x - a),$$

$$zq = -(y - b).$$

Therefore $\qquad\qquad z^2(1 + p^2 + q^2) = 1.$ $\qquad\qquad$ □

**Exercise 1.2.1:**   Eliminate the arbitrary function $F$ from each of the following equations and find the corresponding p.d.e.

(i)   $F(z - xy, x^2 + y^2) = 0.$   **Ans. :**   $py - qx = y^2 - x^2.$

(ii)   $F(xy, x + y - z) = 0.$   **Ans. :**   $px - qy = x - y.$

(iii)   $z = F\left(\dfrac{xy}{z}\right).$   **Ans. :**   $px - qy = 0.$

(iv)   $F(x + y, x - \sqrt{z}) = 0.$   **Ans. :**   $p - q = 2\sqrt{z}.$

(v)   $F\left(\dfrac{xy}{z}, \dfrac{x - y}{z}\right) = 0.$   **Ans. :**   $x^2p + y^2q = z(x + y).$

(vi)   $F(xyz, x + y + z) = 0.$   **Ans. :**   $x(y - z)p + y(z - x)q = z(x - y).$

**Exercise 1.2.2:**   Eliminate the parameters $a$ and $b$ from each of the following equations and find the corresponding p.d.e.

(i)   $z = (x + a)(y + b).$   **Ans. :**   $pq = z.$

(ii)   $2z = (ax + y)^2 + b.$   **Ans. :**   $px + qy = q^2.$

(iii)   $z = ax + by.$   **Ans. :**   $z = px + qy.$

(iv)   $z^2(1 + a^3) = 8(x + ay + b)^3.$   **Ans. :**   $p^3 + q^3 = 27z.$

## Classification of first order p.d.e.

1. **Linear equation:** A first order p.d.e. is said to be a linear equation if it is linear in $p, q$ and $z$, i.e., if it is of the form

$$P(x, y)p + Q(x, y)q = R(x, y)z + S(x, y).$$

   **Example:** $yp - xq = xyz + x.$

2. **Semi-linear equation:** A first order p.d.e. is said to be a semi-linear equation if it is linear in $p$ and $q$ and the coefficients of $p$ and $q$ are functions of $x$ and $y$ only, i.e., if it is of the form

$$P(x, y)p + Q(x, y)q = R(x, y, z).$$

   **Example:** $e^x p - yxq = xz^2.$

3. **Quasi-linear equation**: A first order p.d.e. is said to be a quasi-linear equation if it is linear in $p$ and $q$, i.e., if it is of the form

$$P(x, y, z)p + Q(x, y, z)q = R(x, y, z).$$

**Example**: $(x^2 + z^2)p - xyq = z^3x + y^2$.

4. **Non-linear equation**: Partial differential equations of the form $f(x, y, z, p, q) = 0$ that do not come under the previous three types are said to be non-linear equations.

**Example**: $pq = z$ does not belong to any of the first three types. So it is a non-linear first order p.d.e. Also refer to Example 1.2.6.

**Note**: From Examples 1.2.1 – 1.2.6, we observe that by eliminating arbitrary functions, we always produce quasi-linear partial differential equations only. However, we get both quasi-linear as well as non-linear partial differential equations when we eliminate arbitrary constants.

## 1.3    Classification of Integrals

Let us consider a first order p.d.e.

$$f(x, y, z, p, q) = 0. \tag{1.3.1}$$

Essentially a solution of (1.3.1) in a region $D \subseteq \mathbb{R} \times \mathbb{R}$ is given by $z$ as a continuously differentiable function of $x$ and $y$ for $(x, y) \in D$. Further if one computes $p$ and $q$ from it and substitutes them into (1.3.1), then the equation reduces to an identity in $x$ and $y$. There are different types of solutions (integral surfaces) for the first order p.d.e. (1.3.1).

**Note**: A solution $z = z(x, y)$ when interpreted as a surface in three-dimensional space will be called an integral surface of the partial differential equation.

(a) **Complete integral**: A two-parameter family of solutions

$$z = F(x, y, a, b), \tag{1.3.2}$$

is called a 'complete integral' of (1.3.1), if in the region considered, the rank of the matrix

$$M = \begin{pmatrix} F_a & F_{xa} & F_{ya} \\ F_b & F_{xb} & F_{yb} \end{pmatrix}$$

is two.

(b) **General integral**: In (1.3.2), if we take $b = \phi(a)$, we get a one-parameter family of solutions of (1.3.1), which is a sub-family of the two-parameter family (1.3.2) as

$$z = F(x, y, a, \phi(a)). \tag{1.3.3}$$

The envelope of (1.3.3), if it exists, is obtained by eliminating $a$ between (1.3.3) and

$$F_a + F_b \phi'(a) = 0. \tag{1.3.4}$$

In fact, if (1.3.4) can be solved for $a$, then

$$a = a(x, y).$$

Substituting for $a$ in (1.3.3), we obtain an integral surface (refer to Lemma (1.3.1)) as

$$z = F(x, y, a(x, y), \phi(a(x, y))). \tag{1.3.5}$$

If the function $\phi$, which defines this sub-family is arbitrary, then such a solution is called a general integral (general solution) of (1.3.1). When a particular function $\phi$ is used, we obtain a particular solution of the p.d.e. Different choices of $\phi$ may give different particular solutions of the p.d.e.

**Note**: A general integral hence involves an arbitrary function and the following lemma shows that it is indeed a solution of the given p.d.e.

**Lemma 1.3.1**: Let $z = F(x, y, a)$ be a one-parameter family of solutions of (1.3.1). Then the envelope of this one-parameter family, if it exists, is also a solution of (1.3.1).

**Proof**: Note that the envelope is obtained by eliminating $a$ between

$$z = F(x, y, a), \tag{1.3.6}$$

and $\qquad\qquad\qquad 0 = F_a(x, y, a). \tag{1.3.7}$

Hence the envelope will be given by $z = G(x, y) = F(x, y, a(x, y))$ where $a(x, y)$ is obtained from (1.3.7) by solving for $a$ in terms of $x$ and $y$.

The envelope will satisfy the p.d.e. (1.3.1). For,

$$G_x = F_x + F_a a_x = F_x,$$

$$G_y = F_y + F_a a_y = F_y,$$

since $F_a = 0$. Thus, the envelope will have the same partial derivatives as those of a member of the family. The partial differential equation at every point being only a relation to be satisfied between these derivatives, the envelope satisfies the p.d.e. (1.3.1).    □

(c) **The singular integral**: In addition to the 'general integral', we can sometimes obtain yet another solution by finding the envelope of the two-parameter family (1.3.2). This is obtained by eliminating $a$ and $b$ from the equations

$$z = F(x, y, a, b), \quad F_a = 0, \quad F_b = 0, \tag{1.3.8}$$

and is called the singular integral of (1.3.1).

**Lemma 1.3.2**: The singular integral is also a solution.

**Proof**: Let $z = F(x, y, a, b)$ be a complete integral. We will show that the envelope of this two-parameter family, if it exists, is also a solution.

Note that the envelope is obtained by eliminating $a$ and $b$ between

$$z = F(x, y, a, b), \tag{1.3.9}$$

$$0 = F_a(x, y, a, b), \tag{1.3.10}$$

$$0 = F_b(x, y, a, b). \tag{1.3.11}$$

Hence the envelope will be given by

$$z = G(x, y) = F(x, y, a(x, y), b(x, y)),$$

where $a(x, y)$ and $b(x, y)$ are obtained from (1.3.10) and (1.3.11) by solving for $a$ and $b$ in terms of $x$ and $y$.

The envelope will satisfy the given partial differential equation. For,

$$G_x = F_x + F_a a_x + F_b b_x = F_x,$$
$$G_y = F_y + F_a a_y + F_b b_y = F_y, \quad \text{since } F_a = 0, \; F_b = 0.$$

That is, the envelope will have the same partial derivatives as a member of the family and for the same reasons as in Lemma 1.3.1, this envelope, if it exists, is also a solution.    □

The singular integral can, however, be found from the p.d.e. itself without knowing any complete integral.

**Lemma 1.3.3**: The singular solution is obtained by eliminating $p$ and $q$ from the equations

$$\left.\begin{array}{l} f(x, y, z, p, q) = 0, \\ f_p(x, y, z, p, q) = 0, \\ f_q(x, y, z, p, q) = 0. \end{array}\right\} \tag{1.3.12}$$

**Proof**: Since $z = F(x, y, a, b)$ is a complete integral of (1.3.1), the equation

$$f(x, y, F(x, y, a, b), F_x(x, y, a, b), F_y(x, y, a, b)) = 0, \tag{1.3.13}$$

which holds identically for all $a$ and $b$ can be differentiated with respect to $a$ and $b$, and hence leads to

$$\left.\begin{array}{l} f_z F_a + f_p F_{xa} + f_q F_{ya} = 0, \\ f_z F_b + f_p F_{xb} + f_q F_{yb} = 0. \end{array}\right\} \tag{1.3.14}$$

On the singular integral, $F_a = 0$ and $F_b = 0$. Therefore the equations in (1.3.14) simplify to

$$f_p F_{xa} + f_q F_{ya} = 0,$$

$$f_p F_{xb} + f_q F_{yb} = 0.$$

On this surface, $F_{xa}F_{yb} - F_{xb}F_{ya} \neq 0$ (since $F_a = 0, F_b = 0$) and hence $f_p = 0, f_q = 0$. Otherwise the matrix

$$\begin{pmatrix} F_a & F_{xa} & F_{ya} \\ F_b & F_{xb} & F_{yb} \end{pmatrix}$$

will not have rank two contradicting the fact that $z = F(x, y, a, b)$ is a complete integral. Hence the lemma. $\qquad\square$

**Special integral**: Usually (but not always), the three classes (a), (b), and (c) given previously include all the integrals of the first order p.d.e. (1.3.1). However, there are some solutions of certain first order partial differential equations that do not fall under any of the three classes (a), (b), or (c). Such solutions are called 'Special integrals'.

**Example 1.3.1**:   $F(x + y, x - \sqrt{z}) = 0$ is the general integral of the equation $p - q = 2\sqrt{z}$. But $z = 0$ also satisfies this equation and it cannot be obtained from the general integral. It is a special integral of the equation.
A complete integral of the p.d.e. is

$$\sqrt{z} = \frac{(ax + y)}{(a - 1)} + b. \qquad\square$$

**Example 1.3.2**: Consider

$$f(x, y, z, p, q) = z - px - qy - p^2 - q^2 = 0. \qquad (1.3.15)$$

The two-parameter family of planes

$$z = F(x, y, a, b) = ax + by + a^2 + b^2, \qquad (1.3.16)$$

is a complete integral of (1.3.15), since the matrix

$$\begin{pmatrix} x + 2a & 1 & 0 \\ y + 2b & 0 & 1 \end{pmatrix}$$

is of rank two and these planes satisfy the p.d.e. (1.3.15).
Let us now take $b = \sqrt{(1 - a^2)}$. Then

$$z = F(x, y, a, \sqrt{1 - a^2}) = ax + \sqrt{(1 - a^2)}\, y + 1,$$

$$\frac{\partial F}{\partial a} = x - \frac{ay}{\sqrt{(1 - a^2)}} = 0.$$

On eliminating $a$, we get

$$(z - 1)^2 = (x^2 + y^2).$$

This is a particular solution of the given p.d.e.
If we take $b = a$, then $z = ax + ay + 2a^2$.

$$\frac{\partial F}{\partial a} = 0 \Rightarrow x + y = -4a.$$

On eliminating $a$, the envelope is

$$8z = -(x + y)^2.$$

This is another particular solution of the given p.d.e.
Now from Equation (1.3.16), the conditions $F_a = 0$ and $F_b = 0$ become

$$F_a = (x + 2a) = 0, \qquad (1.3.17)$$

$$F_b = (y + 2b) = 0, \qquad (1.3.18)$$

respectively. On eliminating $a$ and $b$ between Equations (1.3.16), (1.3.17), and (1.3.18), we obtain the singular integral

$$4z = -(x^2 + y^2), \tag{1.3.19}$$

which is a paraboloid of revolution. $\qquad\qquad\qquad\qquad\qquad\qquad\qquad\square$

**Note**: Using Lemma 1.3.3, the singular integral can also be obtained directly by eliminating $p$ and $q$ between (1.3.15) and

$$f_p = -x - 2p = 0, \tag{1.3.20}$$

$$f_q = -y - 2q = 0. \tag{1.3.21}$$

**The Cauchy problem (Initial value problem):**
Given a first order partial differential equation and a curve in space, the Cauchy problem is to find an integral surface (i.e., a solution) of the given p.d.e., which contains the given curve. In other words, given a p.d.e. (not necessarily non-linear)

$$f(x, y, z, p, q) = 0,$$

and a curve $x = x_0(s), y = y_0(s), z = z_0(s), s \in [a, b]$, the Cauchy problem is to find a solution $z = z(x, y)$ of the partial differential equation such that $z_0(s) = z(x_0(s), y_0(s))$ for all $s \in [a, b]$. (Refer to Theorems 1.10.1 and 1.11.1).

**Exercise 1.3.1**: Show that $(x - a)^2 + (y - b)^2 + z^2 = 1$ is a complete integral of $z^2(1 + p^2 + q^2) = 1$. By taking $b = 2a$, show that the envelope of the sub-family is $(y - 2x)^2 + 5z^2 = 5$, which is a particular solution. Show further that $z = \pm 1$ are the singular integrals.

**Exercise 1.3.2**: Show that $z = ax + (y/a) + b$ is a complete integral of $pq = 1$. This problem has no singular integral. Find the particular solution corresponding to the sub-family $b = a$.

**Exercise 1.3.3**: Show that $2z = (ax+y)^2+b$ is a complete integral of $px+qy-q^2 = 0$.

## 1.4 Linear Equations of the First Order

This section describes a method for finding a general integral (general solution) for a quasi-linear equation. To this end, we prove the following theorem.

**Theorem 1.4.1**: The general solution of the quasi-linear equation (or Lagrange's equation)

$$P(x, y, z)p + Q(x, y, z)q = R(x, y, z), \tag{1.4.1}$$

where $P, Q$, and $R$ are continuously differentiable functions of $x, y$, and $z$ (and not vanishing simultaneously) is

$$F(u, v) = 0, \tag{1.4.2}$$

where $F$ is an arbitrary differentiable function of $u$ and $v$ and

$$u(x, y, z) = c_1 \quad \text{and} \quad v(x, y, z) = c_2, \tag{1.4.3}$$

are two independent solutions of the system

$$\frac{dx}{P(x, y, z)} = \frac{dy}{Q(x, y, z)} = \frac{dz}{R(x, y, z)}. \tag{1.4.4}$$

**Note**: $u(x, y, z) = c_1$ is a solution of (1.4.4), if at every point on the surface, the tangent plane at that point contains the line through that point with direction ratios $(P, Q, R)$. Observe that (1.4.4) being a system of ordinary differential equations, its solution is actually a curve. The curve is nothing but the intersection of the two surfaces $u = c_1$ and $v = c_2$ and the line through each point of the curve with direction ratios $(P, Q, R)$ is actually the tangent to the curve at that point. These curves are called the **characteristic curves** of the p.d.e. (1.4.1). We will discuss this more in detail later in Section 1.10. Sometimes Equation (1.4.2) is written as $v = G(u)$, where $G$ is an arbitrary differentiable function of $u$.

**Proof**: Since $u(x, y, z) = c_1$ is a solution of (1.4.4), we have

$$Pu_x + Qu_y + Ru_z = 0. \tag{1.4.5}$$

Similarly, since $v(x, y, z) = c_2$ is a solution of (1.4.4), we also have

$$Pv_x + Qv_y + Rv_z = 0. \tag{1.4.6}$$

Therefore solving these equations for $P, Q$, and $R$, we have

$$\frac{P}{\partial(u, v)/\partial(y, z)} = \frac{Q}{\partial(u, v)/\partial(z, x)} = \frac{R}{\partial(u, v)/\partial(x, y)}. \tag{1.4.7}$$

This is rendered possible because of the fact that $u = c_1$ and $v = c_2$ are independent. We showed earlier (refer to Example 1.2.2) that the relation (1.4.2) leads to the p.d.e.

$$p\frac{\partial(u, v)}{\partial(y, z)} + q\frac{\partial(u, v)}{\partial(z, x)} = \frac{\partial(u, v)}{\partial(x, y)}. \tag{1.4.8}$$

On substituting from (1.4.7) into (1.4.8), we see that (1.4.2) is a solution of the Equation (1.4.1) if $u = c_1$ and $v = c_2$ are independent solutions of (1.4.4).    □

**Example 1.4.1**: Find the general solution of $xp + yq = z$.

**Solution**: The auxiliary equations (Equations 1.4.4) are

$$\frac{dx}{x} = \frac{dy}{y} = \frac{dz}{z}.$$

On integrating these auxiliary equations, we obtain

$$\log x = \log c_1 y \Rightarrow u = \left(\frac{x}{y}\right) = c_1,$$

$$\log y = \log c_2 z \Rightarrow v = \left(\frac{y}{z}\right) = c_2.$$

Therefore the general integral is

$$F\left(\frac{x}{y}, \frac{y}{z}\right) = 0,$$

where $F$ is an arbitrary differentiable function. Another form of the general integral is

$$z = yG\left(\frac{x}{y}\right),$$

where $G$ is an arbitrary differentiable function.    □

**Example 1.4.2**: Find the general integral of $yzp + xzq = xy$.

**Solution**: The auxiliary equations are

$$\frac{dx}{yz} = \frac{dy}{xz} = \frac{dz}{xy},$$

$$\frac{dx}{y} = \frac{dy}{x} \Rightarrow u = x^2 - y^2 = c_1,$$

$$\frac{dy}{z} = \frac{dz}{y} \Rightarrow v = z^2 - y^2 = c_2.$$

Therefore the general integral is

$$F(x^2 - y^2, z^2 - y^2) = 0 \quad \text{or} \quad z^2 = y^2 + G(x^2 - y^2),$$

where $F$ and $G$ are arbitrary differentiable functions.    □

**Example 1.4.3**: Find the general solution of
$$x(y^2 - z^2)p - y(z^2 + x^2)q = (x^2 + y^2)z.$$
**Solution**: The auxiliary equations are

$$\frac{dx}{x(y^2 - z^2)} = -\frac{dy}{y(z^2 + x^2)} = \frac{dz}{z(x^2 + y^2)}.$$

Observe that

$$\frac{xdx + ydy + zdz}{x^2(y^2 - z^2) - y^2(z^2 + x^2) + z^2(x^2 + y^2)} = \frac{dz}{z(x^2 + y^2)},$$

$$\Rightarrow \frac{xdx + ydy + zdz}{0} = \frac{dz}{z(x^2 + y^2)},$$

$$\Rightarrow xdx + ydy + zdz = 0.$$

Therefore $x^2 + y^2 + z^2 = c_1$.
Also,

$$\frac{\frac{dx}{x} - \frac{dy}{y}}{y^2 - z^2 + z^2 + x^2} = \frac{dz}{z(x^2 + y^2)},$$

which implies

$$\frac{dx}{x} - \frac{dy}{y} = \frac{dz}{z}.$$

Therefore $\log\left(\frac{yz}{x}\right) = \log c_2$, i.e., $\frac{yz}{x} = c_2$.
Hence a general solution is

$$z = \frac{x}{y}G(x^2 + y^2 + z^2),$$

or

$$F\left(\frac{yz}{x}, x^2 + y^2 + z^2\right) = 0,$$

where $G$ and $F$ are arbitrary differentiable functions.    □
**Example 1.4.4**: Find the general integral of $z_t + zz_x = 0$ and verify that it satisfies the equation.
**Solution**: The auxiliary equations are

$$\frac{dt}{1} = \frac{dx}{z} = \frac{dz}{0}.$$

The two intermediate integrals are

$$u = z = c_1 , \quad v = x - c_1 t = c_2, \quad \text{i.e.,} \quad v = x - zt = c_2.$$

The general integral is

$$F(z, x - zt) = 0 \text{ or } z(x, t) = \phi(x - zt). \tag{1.4.9}$$

On differentiating (1.4.9) with respect to $t$ and $x$ respectively, we obtain

$$z_t = -\frac{z\phi'}{(1 + t\phi')} , \quad z_x = \frac{\phi'}{(1 + t\phi')} .$$

It can be verified that these expressions satisfy the given equation. So far, $\phi$ is an arbitrary function. We can determine its form if we prescribe $z$ as a function of $x$ at $t = 0$. Suppose $z(x, 0) = -x$, then $\phi(x) = -x$ and from (1.4.9) we get

$$z(x, t) = -\frac{x}{1 - t} .$$

**Note**: The solution becomes unbounded as $t \to 1$.   □

**Exercise 1.4.1**: Find the general integrals of

(i)  $z(xp - yq) = y^2 - x^2$.
   **Ans.** : $F(xy, (x - y)^2 + z^2) = 0$.

(ii)  $y^2 p - xyq = x(z - 2y)$.
   **Ans.** : $F(x^2 + y^2, y(y - z)) = 0$.

(iii)  $(z^2 - 2yz - y^2)p + x(y + z)q = x(y - z)$.
   **Ans.** : $F(x^2 + y^2 + z^2, y^2 - 2yz - z^2) = 0$.

(iv)  $yzp + xzq = x + y$.
   **Ans.** : $F(x^2 - y^2, z^2 - 2(x + y)) = 0$.

(v)  $(y + 1)p + (x + 1)q = z$.
   **Ans.** : $F(x^2 - y^2 + 2x - 2y, z(x - y)) = 0$.

(vi)  $(x^2 + y^2)p + 2xyq = (x + y)z$.
   **Ans.** : $F((x + y)/z, (x^2/y) - y) = 0$.

(vii)  $2x(y + z^2)p + y(2y + z^2)q = z^3$.
   **Ans.**: $F(x/yz, (z^2 - 2y)/yz) = 0$.

(viii) $(x^3 + 3xy^2)p + (y^3 + 3x^2y)q = 2(x^2 + y^2)z$.
   **Ans.**: $xy/z^2 = G((x^2 - y^2)/z)$.

(ix) $x(y - z)p + y(z - x)q = z(x - y)$.
   **Ans.**: $xyz = G(x + y + z)$.

We will now state the theorem for the case when the number of independent variables is more than two. This result would be required in Section 1.6.

**Theorem 1.4.2**:

If $u_i(x_1, \cdots, x_n, z) = c_i$, $(i = 1, 2, \cdots, n)$ are independent solutions of the equations

$$\frac{dx_1}{P_1} = \frac{dx_2}{P_2} = \cdots = \frac{dx_n}{P_n} = \frac{dz}{R},$$

where $P_1, P_2, \cdots P_n$, and $R$ are continuously differentiable functions of $x_1, x_2, \cdots x_n$, and $z$, not simultaneously zero, then the relation $\phi(u_1, u_2, \cdots, u_n) = 0$ where $\phi$ is an arbitrary differentiable function is a general solution of the quasi-linear p.d.e.

$$P_1 \frac{\partial z}{\partial x_1} + P_2 \frac{\partial z}{\partial x_2} + \cdots + P_n \frac{\partial z}{\partial x_n} = R.$$

**Proof**: Similar to the proof given in Theorem 1.4.1.     □

## 1.5   Pfaffian Differential Equations

By a Pfaffian differential equation, we mean a differential equation of the form

$$F_1(x_1, \cdots, x_n)dx_1 + F_2(x_1, \cdots, x_n)dx_2 + \cdots + F_n(x_1, \cdots, x_n)dx_n = 0, \qquad (1.5.1)$$

where $F_i$'s $(i = 1, \cdots n)$ are continuous functions. The expression on the left-hand side of Equation (1.5.1) is called a Pfaffian differential form.

**Definition 1.5.1**: A Pfaffian differential form is said to be exact if we can find a continuously differentiable function $u(x_1, \cdots, x_n)$ such that

$$du = F_1(x_1, \cdots, x_n)dx_1 + F_2(x_1, \cdots, x_n)dx_2 + \cdots + F_n(x_1, \cdots, x_n)dx_n.$$

**Definition 1.5.2**: A Pfaffian differential equation is said to be integrable if there exists a non-zero differentiable function $\mu(x_1, \cdots, x_n)$ such that the Pfaffian differential form

$$\mu \left[ F_1(x_1, \cdots, x_n)dx_1 + \cdots + F_n(x_1, \cdots, x_n)dx_n \right],$$

is exact. The function $\mu(x_1, \cdots, x_n)$ is called the integrating factor and $u(x_1, \cdots, x_n) = c$, where $c$ is an arbitrary constant, is called the integral of the corresponding Pfaffian differential equation.

**Note**: A Pfaffian differential equation (1.5.1) is said to be exact if the Pfaffian differential form on the left-hand side of Equation (1.5.1) is exact.

**Note**: Observe that $u(x_1, \cdots, x_n) = c$ is a surface in $I\!\!R^n$ such that at every point on it the normal has direction ratios $(F_1, F_2, \cdots F_n)$.

**Theorem 1.5.1**: There always exists an integrating factor for a Pfaffian differential equation in two variables.

**Proof**:   Consider $P(x, y)dx + Q(x, y)dy = 0$. If $Q(x, y) \neq 0$, then

$$\frac{dy}{dx} = -\frac{P(x, y)}{Q(x, y)}.$$

From the existence theorem for a first-order ordinary differential equation, the previous equation has a solution $F(x, y) = c_1$. Then

$$\frac{\partial F}{\partial x}dx + \frac{\partial F}{\partial y}dy = 0.$$

This exact Pfaffian form differs from $Pdx + Qdy = 0$ by a factor as they have the same solution. This factor is nothing but the integrating factor.   □

In general, a Pfaffian differential equation in more than two variables may not be integrable. In the next theorem, we shall derive a necessary and sufficient condition for the integrability of a Pfaffian differential equation in three variables. We shall first prove two lemmas that are used in the theorem.

It was shown in Example 1.2.3 that if there is a functional dependence between $u(x, y)$ and $v(x, y)$ not involving $x$ and $y$ explicitly, then $\dfrac{\partial(u, v)}{\partial(x, y)} = 0$. We shall prove the converse in the following lemma.

**Lemma 1.5.1**: Let $u(x, y)$ and $v(x, y)$ be two functions of $x$ and $y$ such that

$$\frac{\partial v}{\partial y} \neq 0. \tag{1.5.2}$$

If, further

$$\frac{\partial(u, v)}{\partial(x, y)} = 0, \tag{1.5.3}$$

then there exists a relation

$$F(u, v) = 0, \tag{1.5.4}$$

between $u$ and $v$ not involving $x$ and $y$ explicitly.

**Proof:** Since $\dfrac{\partial v}{\partial y} \neq 0$, we can eliminate $y$ between $u = u(x, y)$ and $v = v(x, y)$ and obtain the relation

$$F(u, v, x) = 0. \tag{1.5.5}$$

We will now show that $F$ does not depend on $x$, leading to (1.5.4).
On differentiating (1.5.5) with respect to $x$ and $y$, we get, respectively

$$\frac{\partial F}{\partial x} + \frac{\partial F}{\partial u}\frac{\partial u}{\partial x} + \frac{\partial F}{\partial v}\frac{\partial v}{\partial x} = 0, \tag{1.5.6}$$

$$\frac{\partial F}{\partial u}\frac{\partial u}{\partial y} + \frac{\partial F}{\partial v}\frac{\partial v}{\partial y} = 0. \tag{1.5.7}$$

On eliminating $\dfrac{\partial F}{\partial v}$ from these equations, which is possible if $\dfrac{\partial v}{\partial x} \neq 0$, we find that

$$\frac{\partial F}{\partial x}\frac{\partial v}{\partial y} + \frac{\partial(u, v)}{\partial(x, y)}\frac{\partial F}{\partial u} = 0, \tag{1.5.8}$$

i.e.,
$$\frac{\partial F}{\partial x}\frac{\partial v}{\partial y} = 0,$$

by virtue of (1.5.3).

However, since $\dfrac{\partial v}{\partial y} \neq 0$, this implies that $\dfrac{\partial F}{\partial x} = 0$.

Suppose $\dfrac{\partial v}{\partial x} = 0$, then $\dfrac{\partial u}{\partial x} = 0$ (why?). In this case also, Equation (1.5.6) implies that $\dfrac{\partial F}{\partial x} = 0$. Hence $F$ is independent of $x$.     □

**Lemma 1.5.2:**   If $\vec{X} \cdot \mathrm{curl}\vec{X} = 0$ where $\vec{X} = (P, Q, R)$ and $\mu$ is an arbitrary differentiable function of $x, y,$ and $z$, then

$$\mu\vec{X} \cdot \mathrm{curl}(\mu\vec{X}) = 0.$$

**Note:**  By $\vec{X} = (P, Q, R)$, we mean the vector $\vec{X} = P\vec{i} + Q\vec{j} + R\vec{k}$, where $\vec{i}, \vec{j},$ and $\vec{k}$ are the unit vectors in the positive $x, y,$ and $z$  directions respectively.

**Proof**: Consider

$$\mu \vec{X} \cdot \text{curl}(\mu \vec{X}) = \sum_{x,y,z} (\mu P) \left[ \frac{\partial(\mu R)}{\partial y} - \frac{\partial(\mu Q)}{\partial z} \right],$$

$$= \mu^2 \sum_{x,y,z} P \left[ \frac{\partial R}{\partial y} - \frac{\partial Q}{\partial z} \right] - \mu \sum_{x,y,z} \left[ PQ \frac{\partial \mu}{\partial z} - PR \frac{\partial \mu}{\partial y} \right],$$

$$= \mu^2 \sum_{x,y,z} P \left[ \frac{\partial R}{\partial y} - \frac{\partial Q}{\partial z} \right],$$

$$= \mu^2 (\vec{X} \cdot \text{curl} \vec{X}).$$

The lemma follows from the previous result.                    □

Conversely, if $\mu \vec{X} \cdot \text{curl}(\mu \vec{X}) = 0$ then $\vec{X} \cdot \text{curl} \vec{X} = 0$ for $\mu \neq 0$.

**Theorem 1.5.2**: A necessary and sufficient condition that the Pfaffian differential equation

$$\vec{X} \cdot d\vec{r} = P(x, y, z)dx + Q(x, y, z)dy + R(x, y, z)dz = 0, \tag{1.5.9}$$

be integrable is that

$$(\vec{X} \cdot \text{curl} \vec{X}) = 0. \tag{1.5.10}$$

**Proof**: We first show that the condition is necessary. For, if Equation (1.5.9) is integrable, then there exist differentiable functions $\mu(x, y, z)$ and $u(x, y, z)$ such that

$$du = \mu(x, y, z)[P(x, y, z)dx + Q(x, y, z)dy + R(x, y, z)dz], \tag{1.5.11}$$

where $\mu(x, y, z) \neq 0$. However,

$$du = \frac{\partial u}{\partial x}dx + \frac{\partial u}{\partial y}dy + \frac{\partial u}{\partial z}dz . \tag{1.5.12}$$

Comparing Equations (1.5.11) and (1.5.12), we get

$$\mu P = \frac{\partial u}{\partial x} , \quad \mu Q = \frac{\partial u}{\partial y} , \quad \mu R = \frac{\partial u}{\partial z} ,$$

that is, $\mu \vec{X} = \nabla u$. Since $\text{curl}(\nabla u) = 0$, we have $\text{curl}(\mu \vec{X}) = 0$. Hence $\mu \vec{X} \cdot \text{curl}(\mu \vec{X}) = 0$.

Therefore, from the converse of the previous lemma, we have $\vec{X} \cdot \text{curl} \vec{X} = 0$.

Now we shall prove that the condition is sufficient. Suppose $z$ is treated as a constant, the differential equation (1.5.9) becomes

$$P(x, y, z)dx + Q(x, y, z)dy = 0.$$

Since a Pfaffian differential equation in two variables is always integrable, this has a solution of the form $U(x, y, z) = c_1$, where $c_1$ may involve $z$. In addition, there must exist a non-zero differentiable function $\mu$ such that

$$\frac{\partial U}{\partial x} = \mu P, \quad \frac{\partial U}{\partial y} = \mu Q.$$

On multiplying (1.5.9) by $\mu$ and substituting the previous two equations, we get

$$\frac{\partial U}{\partial x}dx + \frac{\partial U}{\partial y}dy + \frac{\partial U}{\partial z}dz + (\mu R - \frac{\partial U}{\partial z})dz = 0.$$

This implies that
$$dU + Kdz = 0, \tag{1.5.13}$$

where $$K = (\mu R - \frac{\partial U}{\partial z}).$$

We are given that $(\vec{X} \cdot \mathrm{curl}\vec{X}) = 0$.

Hence by Lemma 1.5.2, $\mu\vec{X} \cdot \mathrm{curl}(\mu\vec{X}) = 0$. Observe that

$$\mu\vec{X} = (\mu P, \mu Q, \mu R) = (\frac{\partial U}{\partial x}, \frac{\partial U}{\partial y}, \frac{\partial U}{\partial z} + K),$$
$$= (\frac{\partial U}{\partial x}, \frac{\partial U}{\partial y}, \frac{\partial U}{\partial z}) + (0, 0, K),$$
$$= \nabla U + (0, 0, K).$$

Hence

$$\mu\vec{X} \cdot \mathrm{curl}(\mu\vec{X}) = (\frac{\partial U}{\partial x}, \frac{\partial U}{\partial y}, \frac{\partial U}{\partial z} + K) \cdot (\frac{\partial K}{\partial y}, -\frac{\partial K}{\partial x}, 0),$$
$$= \frac{\partial U}{\partial x}\frac{\partial K}{\partial y} - \frac{\partial U}{\partial y}\frac{\partial K}{\partial x}.$$

It follows that
$$\frac{\partial(U, K)}{\partial(x, y)} = 0, \tag{1.5.14}$$

since $\mu\vec{X} \cdot \text{curl}(\mu\vec{X}) = 0$ by Lemma 1.5.2. Hence $K$ can be expressed as a function of $U$ and $z$ alone as Equation (1.5.14) indicates that there is a relation between $U$ and $K$ independent of $x$ and $y$, but not necessarily of $z$ (from Lemma 1.5.1). Therefore Equation (1.5.13) becomes

$$\frac{dU}{dz} + K(U, z) = 0.$$

It possesses a solution $\phi(U, z) = c$, where $c$ is an arbitrary constant. This solution can now be expressed in the form $u(x, y, z) = c$ by using the expression for $U$ in terms of $x, y$, and $z$. Therefore the Pfaffian differential equation (1.5.9) is integrable.    □

**Note**: The Pfaffian differential equation (1.5.9) is, in fact, exact if and only if $\text{curl}\vec{X} = 0$.

**Example 1.5.1**:  Show that the following Pfaffian differential equation is integrable and find its integral

$$ydx + xdy + 2zdz = 0.$$

**Solution**: Observe that $\text{curl}\vec{X} = 0$, where $\vec{X} = (y, x, 2z)$. Therefore the Pfaffian differential equation is exact. In fact, it can be written as

$$ydx + xdy + 2zdz = d(xy + z^2).$$

Hence the integral is

$$u(x, y, z) = xy + z^2 = c.$$    □

**Example 1.5.2**:  Find the integral of $yzdx + 2xzdy - 3xydz = 0$.
**Solution**: Observe that $\vec{X} \cdot \text{curl}\vec{X} = 0$ where $\vec{X} = (yz, 2xz, -3xy)$, and $U = xy^2 = c_1$ is a solution of $ydx + 2xdy = 0$. Then $\mu = \dfrac{y}{z}$. Further,

$$K = \frac{y(-3xy)}{z} = -\frac{3U}{z}.$$

Therefore

$$\frac{dU}{dz} - \frac{3U}{z} = 0,$$

whose solution is $U = cz^3$. Therefore the integral of the given Pfaffian form is

$$u = \frac{xy^2}{z^3} = c.$$    □

**Exercise 1.5.1**: Verify that the Pfaffian differential equations are exact/integrable and find the corresponding integrals.

(i)    $(y^2 + yz)dx + (xz + z^2)dy + (y^2 - xy)dz = 0.$
    **Ans.** : $y(x + z) = c(y + z).$

(ii)    $(1 + yz)dx + x(z - x)dy - (1 + xy)dz = 0.$
    **Ans.** : $(1 + xy)(1 + cy) = (1 + yz).$

(iii)    $yzdx + xzdy + xydz = 0.$
    **Ans.** : $xyz = c.$

(iv)    $yzdx + (x^2y - zx)dy + (x^2z - xy)dz = 0.$
    **Ans.** : $2zy - x(y^2 + z^2) = 2cx.$

(v)    $(6x + yz)dx + (xz - 2y)dy + (xy + 2z)dz = 0.$
    **Ans.** : $3x^2 - y^2 + z^2 + xyz = c.$

(vi)    $z(z - y)dx + z(x + z)dy + x(x + y)dz = 0.$
    **Ans.** : $z(x + y) = c(x + z).$

(vii)    $(2x + y^2 + 2xz)dx + 2xydy + x^2dz = 0.$
    **Ans.** : $x^2 + xy^2 + x^2z = c.$

(viii)    $(1 + yz)dx + z(z - x)dy - (1 + xy)dz = 0.$
    **Ans.** : $(1 + yz)(1 + cy) = (1 + xy).$

# 1.6    Compatible Systems of First Order Partial Differential Equations

**Definition 1.6.1**: The equations

$$f(x, y, z, p, q) = 0, \tag{1.6.1}$$

and

$$g(x, y, z, p, q) = 0, \tag{1.6.2}$$

are compatible on a domain $D$ if

$$\text{(i) } J = \frac{\partial(f, g)}{\partial(p, q)} \neq 0 \text{ on } D, \tag{1.6.3}$$

and
$$\text{(ii) } p = \phi(x, y, z), q = \psi(x, y, z),$$

obtained by solving (1.6.1) and (1.6.2) (in view of (i)), render

$$dz = \phi(x, y, z)dx + \psi(x, y, z)dy, \tag{1.6.4}$$

integrable.

**Theorem 1.6.1**: A necessary and sufficient condition for the integrability of (1.6.4) is

$$[f,g] \equiv \frac{\partial(f,g)}{\partial(x,p)} + p\frac{\partial(f,g)}{\partial(z,p)} + \frac{\partial(f,g)}{\partial(y,q)} + q\frac{\partial(f,g)}{\partial(z,q)} = 0.$$

**Proof**: (1.6.4) is integrable if and only if

$$\vec{X} \cdot \text{curl}\vec{X} = 0, \text{ where } \vec{X} = (\phi, \psi, -1),$$

i.e.,
$$\phi(-\psi_z) + \psi(\phi_z) - (\psi_x - \phi_y) = 0,$$

i.e.,
$$\psi_x + \phi\psi_z = \phi_y + \psi\phi_z. \tag{1.6.5}$$

On substituting $\phi$ and $\psi$ for $p$ and $q$ respectively in (1.6.1) and differentiating it with respect to $x$ and $z$, we obtain

$$f_x + f_p\phi_x + f_q\psi_x = 0,$$

$$f_z + f_p\phi_z + f_q\psi_z = 0.$$

On multiplying the second equation by $\phi$ and adding it to the first, we get

$$f_x + \phi f_z + f_p(\phi_x + \phi\phi_z) + f_q(\psi_x + \phi\psi_z) = 0.$$

Similarly we may deduce from Equation (1.6.2) that

$$g_x + \phi g_z + g_p(\phi_x + \phi\phi_z) + g_q(\psi_x + \phi\psi_z) = 0.$$

Solving these equations, we find that

$$\psi_x + \phi\psi_z = \frac{1}{J}\left\{ \frac{\partial(f,g)}{\partial(x,p)} + \phi\frac{\partial(f,g)}{\partial(z,p)} \right\}. \tag{1.6.6}$$

If we differentiate the given pair of equations with respect to $y$ and $z$, we obtain, after a similar analysis used to obtain Equation (1.6.6)

$$\phi_y + \psi\phi_z = -\frac{1}{J}\left\{ \frac{\partial(f,g)}{\partial(y,q)} + \psi\frac{\partial(f,g)}{\partial(z,q)} \right\}. \tag{1.6.7}$$

On substituting from (1.6.6) and (1.6.7) into (1.6.5) and replacing $\phi$ and $\psi$ by $p$ and $q$ respectively, we see that the condition for integrability of (1.6.4) is that $[f,g] = 0$.
$\square$

**Note**: A solution of (1.6.4) is of the form

$$F(x, y, z, c) = 0, \tag{1.6.8}$$

where $c$ is an arbitrary parameter. Hence we assert that if Equations (1.6.1) and (1.6.2) are compatible then they have a **one-parameter family of common solutions**.

**Example 1.6.1**: Show that the equations

$$f = xp - yq - x = 0, \tag{1.6.9}$$

$$g = x^2p + q - xz = 0, \tag{1.6.10}$$

are compatible and find a one-parameter family of common solutions.

**Solution**: Observe that $\dfrac{\partial(f, g)}{\partial(p, q)} = x(1 + xy) \neq 0$ on the domain $D$ where $D$ does not contain the points $(x, y)$ such that $x = 0$ or $1 + xy = 0$. Then on $D$, we obtain

$$p = \frac{(1 + yz)}{(1 + xy)},$$

$$q = \frac{x(z - x)}{(1 + xy)},$$

and Equation (1.6.4) becomes

$$dz = \frac{1 + yz}{1 + xy}dx + \frac{x(z - x)}{1 + xy}dy. \tag{1.6.11}$$

Then

$$dz - dx = \frac{y(z - x)}{1 + xy}dx + \frac{x(z - x)}{1 + xy}dy,$$

$$\frac{dz - dx}{z - x} = \frac{ydx + xdy}{1 + xy} \Rightarrow z = x + c(1 + xy).$$

Hence (1.6.9) and (1.6.10) are compatible on $D$ as (1.6.11) is integrable and $\dfrac{z - x}{1 + xy} = c$, is a one-parameter family of common solutions. $\square$

**Note**: Compatibility can also be shown by verifying the conditions $\dfrac{\partial(f, g)}{\partial(p, q)} \neq 0$ and $[f, g] = 0$ on $D$.

**Exercise 1.6.1**: Show that the equations

$$f = p^2 + q^2 - 1 = 0, \qquad (1.6.12)$$

$$g = (p^2 + q^2)x - pz = 0, \qquad (1.6.13)$$

are compatible and find the one-parameter family of common solutions.

**Ans.**: $z^2 = x^2 + (y + c)^2$.

**Note**: For the compatibility of $f(x, y, z, p, q) = 0$ and $g(x, y, z, p, q) = 0$ it is not necessary that every solution of $f(x, y, z, p, q) = 0$ be a solution of $g(x, y, z, p, q) = 0$ or vice-versa as is generally believed. This can be seen from the following observations:

(a) In Example 1.6.1, $z = x(y \mid 1)$ is a solution of (1.6.9) and not of (1.6.10).

(b) In Exercise 1.6.1, $z = (x + y)/\sqrt{2}$ is a solution of (1.6.12) and not of (1.6.13).

## 1.7   Charpit's Method

In this section, we present a method to find a complete integral of a first order p.d.e. Let

$$f(x, y, z, p, q) = 0, \qquad (1.7.1)$$

be the partial differential equation whose complete integral is being sought.

A family of partial differential equations

$$g(x, y, z, p, q, a) = 0, \qquad (1.7.2)$$

is said to be a one-parameter family of partial differential equations compatible with (1.7.1) if (1.7.2) is compatible with (1.7.1) for each value of $a$.

Since $f = 0$ and $g = 0$ are compatible, we have $[f, g] = 0$, i.e.,

$$f_p \frac{\partial g}{\partial x} + f_q \frac{\partial g}{\partial y} + (pf_p + qf_q)\frac{\partial g}{\partial z} - (f_x + pf_z)\frac{\partial g}{\partial p} - (f_y + qf_z)\frac{\partial g}{\partial q} = 0. \qquad (1.7.3)$$

This is a quasi-linear first order p.d.e. for $g$ with $x, y, z, p$, and $q$ as the independent variables.

Our problem, i.e., finding a one-parameter family of partial differential equations (1.7.2), which is such that each member of the family is compatible with the given p.d.e. (1.7.1) then is to find a solution of this Equation (1.7.3), in as simple a form as possible, involving $p$ or $q$ or both and an arbitrary constant $a$. This we do by finding an integral of the auxiliary equations (refer to Theorem 1.4.2)

$$\frac{dx}{f_p} = \frac{dy}{f_q} = \frac{dz}{pf_p + qf_q} = -\frac{dp}{f_x + pf_z} = -\frac{dq}{f_y + qf_z}. \qquad (1.7.4)$$

Charpit's method consists of choosing a one-parameter family of partial differential equations (1.7.2) that is such that each member of the family is compatible with the given equation (1.7.1). Then, solving for $p$ and $q$ from (1.7.1) and (1.7.2), we get

$$p = \phi(x, y, z, a), \quad q = \psi(x, y, z, a).$$

Then

$$dz = \phi dx + \psi dy, \qquad (1.7.5)$$

is integrable by virtue of the fact that (1.7.1) and (1.7.2) are compatible. An integral of (1.7.5) will be of the form

$$F(x, y, z, a, b) = 0.$$

As this is a two-parameter family of solutions of Equation (1.7.1), it will be a complete integral of (1.7.1).

**Example 1.7.1**: Find a complete integral of $f = z^2 - pqxy = 0$ by Charpit's method.

**Solution**: The auxiliary equations (refer to Equations (1.7.4)) are

$$\frac{dx}{qxy} = \frac{dy}{pxy} = \frac{dz}{2pqxy} = \frac{dp}{2zp - pqy} = \frac{dq}{2zq - pqx},$$

$$\frac{dz}{2z^2} = \frac{pdx + qdy + xdp + ydq}{2z(px + qy)},$$

or

$$\frac{dz}{z} = \frac{d(xp + yq)}{xp + yq},$$

i.e., $z = a(xp + yq)$, where $a$ is an arbitrary constant. Hence the required one-parameter family of p.d.e., which is compatible with $f = 0$ is $g(x, y, z, p, q, a) = z - a(xp + yq) = 0$.

Further, solving for $p$ and $q$ from $f = 0$ and $g = 0$, we obtain $p = \dfrac{z}{cx}$ and $q = \dfrac{cz}{y}$,

where $a(c + \dfrac{1}{c}) = 1$. Now Equation (1.7.5) in this case becomes

$$dz = (\frac{1}{c}\frac{dx}{x} + c\frac{dy}{y})z.$$

On integrating, we obtain $z = bx^{\frac{1}{c}}y^c$. Hence

$$F(x, y, z, b, c) = z - bx^{\frac{1}{c}}y^c = 0,$$

is a complete integral of $f = 0$.
It can be easily verified that the matrix

$$\begin{pmatrix} F_b & F_{bx} & F_{by} \\ F_c & F_{cx} & F_{cy} \end{pmatrix}$$

is of rank two.    □

**Example 1.7.2**: Find a complete integral of

$$f = (p^2 + q^2)y - qz = 0.$$

**Solution**: The auxiliary equations are

$$\frac{dx}{2py} = \frac{dy}{2qy - z} = \frac{dz}{2(p^2 + q^2)y - qz} = \frac{dp}{pq} = \frac{dq}{-p^2},$$

$$\frac{dp}{pq} = \frac{dq}{-p^2} \Rightarrow p^2 + q^2 = a^2,$$

i.e.,    $g(x, y, z, p, q, a) = p^2 + q^2 - a^2 = 0.$

Solving for $p$ and $q$, we get $q = \dfrac{a^2y}{z}$, $p^2 = a^2 - \dfrac{a^4y^2}{z^2}$.

Hence (1.7.5) becomes

$$dz = \frac{a\sqrt{z^2 - a^2y^2}}{z}dx + \frac{a^2y}{z}dy.$$

Hence it follows that $\dfrac{zdz - a^2ydy}{\sqrt{z^2 - a^2y^2}} = adx$. This implies that

$$z^2 - a^2y^2 = (ax + b)^2.$$

Hence the required complete integral is

$$z^2 = a^2y^2 + (ax + b)^2.$$    □

**Example 1.7.3**: Find a complete integral of

$$f = xpq + yq^2 - 1 = 0.$$

**Solution**: The auxiliary equations (1.7.4) are

$$\frac{dx}{xq} = \frac{dy}{xp + 2yq} = \frac{dz}{2(xpq + yq^2)} = \frac{dp}{-pq} = \frac{dq}{-q^2} \ .$$

$$\frac{dp}{p} = \frac{dq}{q} \Rightarrow p = aq \ .$$

Therefore $g = p - aq = 0$. Solving $f = 0$ and $g = 0$ for $p$ and $q$ we get

$$q = \frac{1}{\sqrt{ax + y}} \quad \text{and} \quad p = \frac{a}{\sqrt{ax + y}} \ .$$

Then Equation (1.7.5) becomes $dz = \dfrac{adx + dy}{\sqrt{ax + y}}$. This implies that

$$(z + b)^2 = 4(ax + y),$$

is the required complete integral.    □

**Example 1.7.4**:  Find a complete integral of the p.d.e.

$$f = x^2p^2 + y^2q^2 - 4 = 0.$$

**Solution**: The auxiliary equations are

$$\frac{dx}{2x^2p} = \frac{dy}{2y^2q} = \frac{dz}{2(x^2p^2 + y^2q^2)} = \frac{dp}{-2xp^2} = \frac{dq}{-2yq^2} \ .$$

Suppose we consider the equation

$$\frac{dy}{2y^2q} = \frac{dq}{-2yq^2}.$$

Then

$$\frac{dy}{y} = \frac{dq}{-q}.$$

Therefore $g = qy - a = 0$. Hence

$$q = \frac{a}{y}, \quad p = \frac{\sqrt{4-a^2}}{x} \Rightarrow dz = \frac{\sqrt{4-a^2}}{x}dx + \frac{a}{y}dy.$$

Therefore

$$z = \sqrt{4-a^2}\log x + a\log y + b,$$

is a complete integral of $f = 0$.

Instead, if we consider

$$\frac{dx}{2x^2p} = \frac{dp}{-2xp^2},$$

then $g = xp - a = 0$ and $p = \frac{a}{x}$, $q = \frac{\sqrt{4-a^2}}{y}$. Then Equation (1.7.5) becomes

$$dz = \frac{a}{x}dx + \frac{\sqrt{4-a^2}}{y}dy.$$

Therefore

$$z = a\log x + \sqrt{4-a^2}\log y + b,$$

is another complete integral of $f = 0$.                                     □

**Note**: A first order p.d.e. can have several complete integrals.

**Example 1.7.5**:  Consider the p.d.e.

$$z^2(1 + p^2 + q^2) = 1.$$

The previous equation has

$$(x-a)^2 + (y-b)^2 + z^2 = 1, \tag{1.7.6}$$

and

$$(y - mx - c)^2 = (1+m^2)(1-z^2), \tag{1.7.7}$$

as its complete integrals.

Observe that these two complete integrals are not equivalent, i.e., we cannot obtain one from another merely by a change in the choice of arbitrary constants. In fact, these two families are entirely different types of surfaces. Note that the two complete integrals in Example 1.7.4 are, however, equivalent.

When, however, one complete integral has been obtained, every other solution, including every other complete integral can be obtained. (We shall explain a procedure for this in Section 1.9.) Consider a sub-family of (1.7.6) by putting $b = ma + c$ where $m$ and $c$ are fixed. Its envelope, which is found by eliminating $a$ between

$$(x - a)^2 + (y - ma - c)^2 + z^2 = 1,$$

and

$$(x - a) + (y - ma - c)m = 0,$$

gives the other complete integral

$$(y - mx - c)^2 = (1 + m^2)(1 - z^2). \qquad \square$$

### Some Standard types

**Type I:** $f(p, q) = 0$ (i.e., the given p.d.e. does not involve $x, y$, and $z$ explicitly). The auxiliary equations are

$$\frac{dx}{f_p} = \frac{dy}{f_q} = \frac{dz}{pf_p + qf_q} = \frac{dp}{0} = \frac{dq}{0} .$$

Solving the last equation, we get either $p = a$ (or $q = a$).
Then we solve $f(a, q) = 0$ (or $f(p, a) = 0$) for $q = Q(a)$ (or $p = P(a)$) .
Then

$$dz = adx + Q(a)dy \Rightarrow z = ax + Q(a)y + b ,$$

$$\text{or } (dz = P(a)dx + ady \Rightarrow z = P(a)x + ay + b.)$$

**Example 1.7.6:** Find a complete integral of $f(p, q) = p + q - pq = 0$.
**Solution:** Put $q = a$. Then $p = \dfrac{a}{a - 1}$. The equation (1.7.5) becomes

$$dz = \frac{a}{a - 1}x + ady,$$

which implies that

$$z = \frac{ax}{a - 1} + ay + b,$$

which is the required complete integral.    □

**Type II**: $f(z, p, q) = 0$ (i.e., the given p.d.e. does not involve $x$ and $y$ explicitly).
The auxiliary equations are

$$\frac{dx}{f_p} = \frac{dy}{f_q} = \frac{dz}{pf_p + qf_q} = \frac{dp}{-pf_z} = \frac{dq}{-qf_z}.$$

$$\frac{dp}{p} = \frac{dq}{q} \Rightarrow p = aq.$$

Therefore

$$f(z, aq, q) = 0 \ \text{ or } \ q = Q(a, z) \text{ and also } p = aQ(a, z),$$

$$dz = pdx + qdy = Q(a, z)(adx + dy).$$

The complete integral is

$$\int \frac{dz}{Q(a, z)} = ax + y + b.$$

**Example 1.7.7**: Find a complete integral of the equation

$$zpq - p - q = 0.$$

**Solution**:   Putting $p = aq$ in the previous equation, we get

$$zaq - a - 1 = 0, \quad \text{or } \big[p = 0, q = 0 \text{ in which case } z = \text{constant}\big].$$

If $zaq - a - 1 = 0$, then

$$q = \frac{1 + a}{az} \text{ and } p = \frac{1 + a}{z}.$$

Now $dz = pdx + qdy = \dfrac{1 + a}{z}(dx + \dfrac{1}{a}dy)$, and hence

$$z^2 = \frac{2(1 + a)}{a}(ax + y) + b.$$    □

**Type III**: $g(x, p) = h(y, q)$   (separable type).
The auxiliary equations are

$$\frac{dx}{g_p} = \frac{dy}{-h_q} = \frac{dz}{pg_p - qh_q} = \frac{dp}{-g_x} = \frac{dq}{h_y}.$$

Note that $g_x dx + g_p dp = 0$.
Therefore $g(x, p) = a$.

Hence $h(y, q) = a$.

From these we can solve for $p$ and $q$ as

$$p = G(a, x) \text{ and } q = H(a, y).$$

Then $dz = pdx + qdy$ implies

$$z = \int G(a, x)dx + \int H(a, y)dy + b.$$

**Example 1.7.8**: Find a complete integral of

$$p^2 + q^2 = x + y.$$

**Solution**: Observe that the given equation can be put in the form $p^2 - x = -(q^2 - y) = a$. Then

$$p = \pm\sqrt{x + a} \text{ and } q = \pm\sqrt{y - a}, \text{ so that}$$
$$dz = (\sqrt{x + a})dx + (\sqrt{y - a})dy.$$

Hence
$$z = \frac{2}{3}(x + a)^{3/2} + \frac{2}{3}(y - a)^{3/2} + b. \qquad \square$$

**Type IV** (Clairaut Form): $z = px + qy + g(p, q)$.

Here a complete integral is given by

$$z = ax + by + g(a, b),$$

for, it is a solution and the matrix

$$\begin{pmatrix} x + g_a & 1 & 0 \\ y + g_b & 0 & 1 \end{pmatrix}$$

is of rank two.

**Example 1.7.9**: Find a complete integral of the p.d.e

$$z = px + qy + \log pq.$$

**Solution**: The complete integral is

$$z = ax + by + \log ab. \qquad \square$$

**Example 1.7.10**: Find a complete integral of the p.d.e.

$$pqz = p^2(xq + p^2) + q^2(yp + q^2),$$

or
$$z = xp + yq + \frac{p^4 + q^4}{pq}.$$

**Solution**: The complete integral is

$$z = ax + by + \frac{a^4 + b^4}{ab}. \qquad \square$$

**Example 1.7.11**: Find a complete integral of the p.d.e

$$z = px + qy + pq.$$

**Solution**: We have

$$\frac{dp}{0} = \frac{dq}{0} \Rightarrow p = a.$$

Solving for $q$, we get $q = \dfrac{z - ax}{y + a}$ . Hence

$$dz = adx + \frac{z - ax}{y + a} dy,$$

$$\frac{dz - adx}{z - ax} = \frac{dy}{y + a},$$

which implies that $z - ax = b(y + a)$, i.e.,

$$z = ax + by + ab. \qquad \square$$

**Exercise 1.7.1**: Find complete integrals of the following partial differential equations

(i) $p^2q^2 + x^2y^2 = x^2q^2(x^2 + y^2)$.
  **Ans.** : $z = \frac{1}{3}(x^2 + a^2)^{3/2} + (y^2 - a^2)^{1/2} + b.$

(ii) $px^5 - 4q^3x^2 + 6x^2z - 2 = 0$.
  **Ans.** : $z = \frac{2}{3}(y + a)^{3/2} + be^{3/x^2} + \frac{1}{9} + \frac{1}{3x^2}.$

(iii) $2(z + xp + yq) = yp^2$.
  **Ans.** : $z = \dfrac{ax}{y^2} + \dfrac{b}{y} - \dfrac{a^2}{4y^3}.$

(iv) $p^2 x + q^2 y = z.$
    **Ans. :** $[(1 + a)z]^{1/2} = (ax)^{1/2} + y^{1/2} + b.$

(v) $2z + p^2 + qy + 2y^2 = 0.$
    **Ans. :** $y^2(2z + (a - x)^2 + y^2) = b.$

(vi) $pxy + pq + qy = yz.$
    **Ans. :** $z = ax + be^y(y + a)^{-a}.$

(vii) $2x(z^2 q^2 + 1) = pz.$
    **Ans. :** $z^2 = 2(a^2 + 1)x^2 + 2ay + b.$

(viii) $zpq = p^2 q(x + q) + pq^2(y + p).$
    **Ans. :** $z = ax + by + 2ab.$

(ix) $z(p^2 + q^2) + px + qy = 0.$
    **Ans. :** $z^2 + (ax + y)^2/(1 + a^2) = b.$

(x) $z + xp - x^2 yq^2 - x^3 pq = 0.$
    **Ans. :** $xz = ay + b(1 - ax).$

(xi) $z^2(p^2 z^2 + q^2) = 1.$
    **Ans. :** $ax + y = \dfrac{(a^2 z^2 + 1)^{3/2}}{3a^2} + b.$

## 1.8   Jacobi's Method

Let us consider the following p.d.e.

$$f(x, y, z, u_x, u_y, u_z) = 0. \tag{1.8.1}$$

Here $x, y,$ and $z$ are the independent variables and the dependent variable $u$ does not appear in the equation.

A function $u = F(x, y, z, a, b, c)$ is said to be a complete integral of (1.8.1) if it satisfies the p.d.e. and the associated matrix

$$\begin{pmatrix} F_a & F_{ax} & F_{ay} & F_{az} \\ F_b & F_{bx} & F_{by} & F_{bz} \\ F_c & F_{cx} & F_{cy} & F_{cz} \end{pmatrix}$$

is of rank three.

**Jacobi's method**: For a given p.d.e. of the type (1.8.1), we consider two one-parameter families of partial differential equations

$$h_1(x, y, z, u_x, u_y, u_z, a) = 0, \tag{1.8.2}$$

$$h_2(x, y, z, u_x, u_y, u_z, b) = 0, \tag{1.8.3}$$

such that

$$\frac{\partial(f, h_1, h_2)}{\partial(u_x, u_y, u_z)} \neq 0, \tag{1.8.4}$$

and the Pfaffian form

$$du = u_x dx + u_y dy + u_z dz, \tag{1.8.5}$$

is integrable, where $u_x(x, y, z, a, b)$, $u_y(x, y, z, a, b)$, and $u_z(x, y, z, a, b)$ are obtained by solving (1.8.1), (1.8.2), and (1.8.3) by virtue of (1.8.4). Such $h_1 = 0$ and $h_2 = 0$ are said to be compatible with $f = 0$.

Observe that (1.8.5) is either exact or not integrable at all (why?). The conditions for (1.8.5) to be exact are

$$\frac{\partial u_x}{\partial y} = \frac{\partial u_y}{\partial x}, \quad \frac{\partial u_y}{\partial z} = \frac{\partial u_z}{\partial y}, \quad \frac{\partial u_z}{\partial x} = \frac{\partial u_x}{\partial z}. \tag{1.8.6}$$

The integral of (1.8.5) gives the complete integral of (1.8.1).

**Theorem 1.8.1**: If $h_1 = 0$ and $h_2 = 0$ are compatible with $f = 0$, then $h_1$ and $h_2$ satisfy

$$\frac{\partial(f, h)}{\partial(x, u_x)} + \frac{\partial(f, h)}{\partial(y, u_y)} + \frac{\partial(f, h)}{\partial(z, u_z)} = 0, \tag{1.8.7}$$

where $h = h_i (i = 1, 2)$.

**Proof**: On differentiating (1.8.1) with respect to $x, y,$ and $z$, we get

$$\frac{\partial f}{\partial x} + \frac{\partial f}{\partial u_x} u_{xx} + \frac{\partial f}{\partial u_y} u_{yx} + \frac{\partial f}{\partial u_z} u_{zx} = 0, \tag{1.8.8}$$

$$\frac{\partial f}{\partial y} + \frac{\partial f}{\partial u_x} u_{xy} + \frac{\partial f}{\partial u_y} u_{yy} + \frac{\partial f}{\partial u_z} u_{zy} = 0, \tag{1.8.9}$$

$$\frac{\partial f}{\partial z} + \frac{\partial f}{\partial u_x} u_{xz} + \frac{\partial f}{\partial u_y} u_{yz} + \frac{\partial f}{\partial u_z} u_{zz} = 0. \tag{1.8.10}$$

Consider

$$h(x, y, z, u_x, u_y, u_z) = 0, \tag{1.8.11}$$

where $h = h_i$ $(i = 1, 2)$.

On differentiating (1.8.11) with respect to $x, y,$ and $z$, we obtain

$$\frac{\partial h}{\partial x} + \frac{\partial h}{\partial u_x} u_{xx} + \frac{\partial h}{\partial u_y} u_{yx} + \frac{\partial h}{\partial u_z} u_{zx} = 0, \tag{1.8.12}$$

$$\frac{\partial h}{\partial y} + \frac{\partial h}{\partial u_x} u_{xy} + \frac{\partial h}{\partial u_y} u_{yy} + \frac{\partial h}{\partial u_z} u_{zy} = 0, \tag{1.8.13}$$

$$\frac{\partial h}{\partial z} + \frac{\partial h}{\partial u_x} u_{xz} + \frac{\partial h}{\partial u_y} u_{yz} + \frac{\partial h}{\partial u_z} u_{zz} = 0. \tag{1.8.14}$$

On multiplying (1.8.8) with $\dfrac{\partial h}{\partial u_x}$ and (1.8.12) by $\dfrac{\partial f}{\partial u_x}$ and subtracting, we get

$$\frac{\partial f}{\partial x}\frac{\partial h}{\partial u_x} - \frac{\partial f}{\partial u_x}\frac{\partial h}{\partial x} + (\frac{\partial f}{\partial u_y}\frac{\partial h}{\partial u_x} - \frac{\partial f}{\partial u_x}\frac{\partial h}{\partial u_y})u_{xy} + (\frac{\partial f}{\partial u_z}\frac{\partial h}{\partial u_x} - \frac{\partial f}{\partial u_x}\frac{\partial h}{\partial u_z})u_{xz} = 0, \tag{1.8.15}$$

since $u_{xy} = u_{yx}, u_{yz} = u_{zy}, u_{xz} = u_{zx}$ from (1.8.6).

On multiplying (1.8.9) with $\dfrac{\partial h}{\partial u_y}$ and (1.8.13) by $\dfrac{\partial f}{\partial u_y}$ and subtracting, we get

$$\frac{\partial f}{\partial y}\frac{\partial h}{\partial u_y} - \frac{\partial f}{\partial u_y}\frac{\partial h}{\partial y} + (\frac{\partial f}{\partial u_x}\frac{\partial h}{\partial u_y} - \frac{\partial f}{\partial u_y}\frac{\partial h}{\partial u_x})u_{xy} + (\frac{\partial f}{\partial u_z}\frac{\partial h}{\partial u_y} - \frac{\partial f}{\partial u_y}\frac{\partial h}{\partial u_z})u_{yz} = 0. \tag{1.8.16}$$

On multiplying (1.8.10) with $\dfrac{\partial h}{\partial u_z}$ and (1.8.14) by $\dfrac{\partial f}{\partial u_z}$ and subtracting, we get

$$\frac{\partial f}{\partial z}\frac{\partial h}{\partial u_z} - \frac{\partial f}{\partial u_z}\frac{\partial h}{\partial z} + (\frac{\partial f}{\partial u_x}\frac{\partial h}{\partial u_z} - \frac{\partial f}{\partial u_z}\frac{\partial h}{\partial u_x})u_{xz} + (\frac{\partial f}{\partial u_y}\frac{\partial h}{\partial u_z} - \frac{\partial f}{\partial u_z}\frac{\partial h}{\partial u_y})u_{yz} = 0. \tag{1.8.17}$$

On adding (1.8.15), (1.8.16), and (1.8.17), we get

$$\frac{\partial(f, h)}{\partial(x, u_x)} + \frac{\partial(f, h)}{\partial(y, u_y)} + \frac{\partial(f, h)}{\partial(z, u_z)} = 0,$$

which is the required result.                                                                    $\square$

The previous equation can also be rewritten as

$$f_{u_x}\frac{\partial h}{\partial x} + f_{u_y}\frac{\partial h}{\partial y} + f_{u_z}\frac{\partial h}{\partial z} - f_x\frac{\partial h}{\partial u_x} - f_y\frac{\partial h}{\partial u_y} - f_z\frac{\partial h}{\partial u_z} = 0, \tag{1.8.18}$$

which is a semi-linear p.d.e. for $h$. That is, for a given $f$ if $h_1$ and $h_2$ satisfy (1.8.4) and (1.8.18), then $h_i = 0$ $(i = 1, 2)$ are compatible with $f = 0$. The following example illustrates Jacobi's method.

**Example 1.8.1**: By Jacobi's method, solve the equation

$$z^2 + z u_z - u_x^2 - u_y^2 = 0.$$

**Solution**: This equation is of the form (1.8.1).
The equation for $h_1$ and $h_2$ (i.e., (1.8.18)) in this example is

$$-2u_x \frac{\partial h}{\partial x} - 2u_y \frac{\partial h}{\partial y} + z \frac{\partial h}{\partial z} - (2z + u_z) \frac{\partial h}{\partial u_z} = 0.$$

The auxiliary equations (refer to Theorem 1.4.2) are

$$\frac{dx}{-2u_x} = \frac{dy}{-2u_y} = \frac{dz}{z} = \frac{du_x}{0} = \frac{du_y}{0} = \frac{du_z}{-2z - u_z}.$$

The two independent solutions of the previous system of equations are $u_x = a$ and $u_y = b$, and then from the given equation

$$u_z = \frac{a^2 + b^2 - z^2}{z}.$$

Now $du = adx + bdy + [(a^2 + b^2 - z^2)/z]dz$ is exact. On integrating, we obtain

$$u = ax + by + (a^2 + b^2) \log z - \frac{1}{2}z^2 + c,$$

which is a complete integral of the given equation.                              □

**Example 1.8.2**: Show that a complete integral of $f(u_x, u_y, u_z) = 0$ is $u = ax + by + cz + d$ where $f(a, b, c) = 0$. Hence find the complete integral of $u_x + u_y + u_z - u_x u_y u_z = 0$.

**Solution**: The auxiliary equations are

$$\frac{dx}{f_{u_x}} = \frac{dy}{f_{u_y}} = \frac{dz}{f_{u_z}} = \frac{du_x}{0} = \frac{du_y}{0} = \frac{du_z}{0}.$$

Therefore $u_x = a, u_y = b, u_z = c$, and

$$u = ax + by + cz + d, \tag{1.8.19}$$

where $a, b, c$, and $d$ are constants, will satisfy the given equation if

$$f(a, b, c) = 0. \tag{1.8.20}$$

This equation determines $c$ in terms of $a$ and $b$. Hence (1.8.19) has three arbitrary constants $a, b$, and $d$, and it is a complete integral. For example, a complete integral of $u_x + u_y + u_z - u_x u_y u_z = 0$ is (1.8.19) if

$$a + b + c - abc = 0,$$

i.e., $c = (a + b)/(ab - 1)$.                                                               $\square$

**Exercise 1.8.1**: Solve the following equations by Jacobi's method.

(i)  $z + 2u_z - (u_x + u_y)^2 = 0,$

   **Ans. :**  $u = ax + by + \dfrac{(a + b)^2 z}{2} - \dfrac{z^2}{4} + c.$

(ii)  $u_x x^2 - u_y^2 - \alpha u_z^2 = 0,$

   **Ans. :**  $u = -\dfrac{(a^2 + \alpha b^2)}{x} + ay + bz + c,$

   where $\alpha$ is a fixed constant.

(iii)  $u_x^2 + u_y^2 + u_z = 1,$

   **Ans. :**  $u = ax + by + (1 - a^2 - b^2)z + c.$

(iv)  $xu_x + yu_y = u_z^2,$

   **Ans. :**  $u = b \log x + (a^2 - b) \log y + az + c.$

where $a, b$, and $c$ are arbitrary constants.

We will now apply Jacobi's method to find a complete integral for a first order p.d.e. in two independent variables. Consider the following p.d.e.

$$f(x, y, z, p, q) = 0. \tag{1.8.21}$$

The solution of (1.8.21) is a relation between $x, y$, and $z$. If this relation is $u(x, y, z) = c$, then

$$p = -(\frac{\partial u}{\partial x})/(\frac{\partial u}{\partial z}) = -\frac{u_x}{u_z},$$

$$q = -(\frac{\partial u}{\partial y})/(\frac{\partial u}{\partial z}) = -\frac{u_y}{u_z}.$$

On substituting the previous in (1.8.21), we obtain

$$g(x, y, z, u_x, u_y, u_z) = 0, \tag{1.8.22}$$

which is in the form given in (1.8.1) and can be solved by Jacobi's method discussed earlier, which yields

$$u = F(x, y, z, a, b) + c.$$

In this, if we choose $u = c$, we get a complete integral of (1.8.21) as $F(x, y, z, a, b) = 0$.

**Example 1.8.3**: Find a complete integral of the equation $p^2 x + q^2 y = z$ by Jacobi's method.

**Solution**:

$$p = -\frac{u_x}{u_z}, \quad q = -\frac{u_y}{u_z}.$$

The given equation becomes

$$x u_x^2 + y u_y^2 - z u_z^2 = 0.$$

The auxiliary equations are

$$\frac{dx}{2xu_x} = \frac{dy}{2yu_y} = \frac{dz}{-2zu_z} = \frac{du_x}{-u_x^2} = \frac{du_y}{-u_y^2} = \frac{du_z}{u_z^2}.$$

The two solutions of these equations are

$$x u_x^2 = a, \quad y u_y^2 = b,$$

whence $u_x = (a/x)^{1/2}, u_y = (b/y)^{1/2}$ and then $u_z = [(a + b)/z]^{1/2}$.

On integrating

$$du = (\frac{a}{x})^{1/2} dx + (\frac{b}{y})^{1/2} dy + (\frac{a+b}{z})^{1/2} dz,$$

we obtain

$$u = 2(ax)^{1/2} + 2(by)^{1/2} + 2((a + b)z)^{1/2} + c.$$

Writing $u = c$, we get the complete integral of the given p.d.e. as

$$z = \left[ \left( \frac{ax}{a+b} \right)^{1/2} + \left( \frac{by}{a+b} \right)^{1/2} \right]^2. \qquad \square$$

**Exercise 1.8.2**: Solve the following equations by Jacobi's method.

(i)  $(p^2 + q^2)y = qz$.  (iv)  $p^2 z + q^2 = 4$.

(ii)  $p = (z + qy)^2$.    (v)   $q - px - p^2 = 0$.

(iii) $z^3 = pqxy$.

# 1.9    Integral Surfaces Through a Given Curve: The Cauchy Problem

This section deals with finding an integral surface passing through a given curve. Let us first discuss the problem for a quasi-linear p.d.e. Here we will obtain the required integral surface from the general integral.

Let $F(u, v) = 0$ (or $v = G(u)$) be the general integral of the partial differential equation $Pp + Qq = R$, where $F$ is an arbitrary function of $u$ and $v$ (or $G$ is an arbitrary function of $u$), where $u(x, y, z) = c_1$, $v(x, y, z) = c_2$ are the solutions of Equation (1.4.4).

Let $C$ be the given curve whose parametric equations are given by

$$x = x_0(s), y = y_0(s), z = z_0(s),$$

where $s$ is a parameter (not necessarily the arc length from a fixed point). We want to find a particular $F$ such that the surface $F(u, v) = 0$ contains the given curve $C$. This is done as follows. Consider the equations

$$u(x, y, z) = c_1, \quad v(x, y, z) = c_2.$$

Substituting $x = x_0(s), y = y_0(s), z = z_0(s)$ in these equations, we get $u(x_0(s), y_0(s), z_0(s)) = c_1$ and $v(x_0(s), y_0(s), z_0(s)) = c_2$. Eliminating $s$ between them, we get a relation between $c_1$ and $c_2$. Let the relation between them be given by $F(c_1, c_2) = 0$. Then $F(u, v) = 0$ is the required solution as $F(u, v) = 0$ is an integral surface. Moreover the initial curve lies on it. For,

$$F(u(x_0(s), y_0(s), z_0(s)), v(x_0(s), y_0(s), z_0(s))) = F(c_1, c_2) = 0.$$

Sometimes the solution can also be obtained by assuming $v = G(u)$ and determining G.

Suppose we assume the form of the integral surface containing the given curve $C$ as $v = G(u)$. Since the given curve $C$ lies on the surface, we find that substituting $x = x_0(s), y = y_0(s), z = z_0(s)$ in the relation $v(x, y, z) = G(u(x, y, z))$ may sometimes enable us to find the explicit form of the function $G$ as $G = G(s)$ as illustrated in the following example. However, it may not always be possible to do so.

**Example 1.9.1**: Find the integral surface of the equation

$$(2xy - 1)p + (z - 2x^2)q = 2(x - yz),$$

which passes through the line $x_0(s) = 1, y_0(s) = 0$, and $z_0(s) = s$.

**Solution**: The auxiliary equations (1.4.4) are

$$\frac{dx}{2xy - 1} = \frac{dy}{z - 2x^2} = \frac{dz}{2(x - yz)},$$

$$= \frac{zdx + dy + xdz}{0},$$

$$= \frac{2xdx + 2ydy + dz}{0}.$$

Therefore  $u = y + xz = c_1, \ v = x^2 + y^2 + z = c_2$.

The general integral is

$$x^2 + y^2 + z = G(y + xz),$$

where $G$ is arbitrary. We want to choose $G$ such that the given curve lies on the surface $x^2 + y^2 + z = G(y + xz)$. This happens if $G(s) = 1 + s$. Therefore the required integral surface is

$$x^2 + y^2 + z = G(y + xz) = 1 + y + xz,$$

i.e.,

$$x^2 + y^2 - xz - y + z = 1. \qquad \square$$

**Example 1.9.2**: Find the integral surface of the equation

$$x^3 p + y(3x^2 + y)q = z(2x^2 + y),$$

which passes through the curve $x_0 = 1, y_0 = s, z_0 = s(1 + s)$.

**Solution**: The auxiliary equations are

$$\frac{dx}{x^3} = \frac{dy}{y(3x^2 + y)} = \frac{dz}{z(2x^2 + y)} = \frac{-x^{-1}dx + y^{-1}dy - z^{-1}dz}{0}.$$

Therefore

$$-\frac{dx}{x} + \frac{dy}{y} - \frac{dz}{z} = 0 \Rightarrow u = \frac{y}{xz} = c_1.$$

$$\frac{dx}{x^3} = \frac{dy}{y(3x^2 + y)} \Rightarrow \frac{(3x^2 + y)dx}{x^3} = \frac{dy}{y} = \frac{(3x^2 + y)dx + dy}{x^3 + y}.$$

Hence

$$\frac{(3x^2 + y)dx + dy + xdy}{x^3 + y + xy} = \frac{dy}{y} \Rightarrow v = \frac{(x^3 + y + xy)}{y} = c_2 .$$

On substituting $x = 1$, $y = s$, $z = s(1 + s)$ in $u = c_1$ and $v = c_2$, we get $\dfrac{1}{s + 1} = c_1$ and $\dfrac{1 + 2s}{s} = c_2$. Eliminating $s$ between them gives $F(c_1, c_2) = c_1 c_2 - c_1 - c_2 + 2 = 0$. Therefore $F(u, v) = 0$ implies that

$$\begin{aligned} yz &= x^3 z + xyz - x^2 y - y^2, \\ &= (x^2 + y)(xz - y), \end{aligned}$$

which is the required integral surface.                                        □

Let us now discuss the same problem for a non-linear p.d.e. Here we will obtain the solution from a complete integral.

Let

$$F(x, y, z, a, b) = 0, \tag{1.9.1}$$

be a complete integral of

$$f(x, y, z, p, q) = 0. \tag{1.9.2}$$

We are interested in finding a solution of $(1.9.2)$ that passes through the curve $C \colon x = x_0(s), y = y_0(s), z = z_0(s)$, $s$ being a parameter.

We expect this solution to be an envelope of a one-parameter sub-family of $(1.9.1)$. Let $E$ be this envelope that contains the curve $C$. Let $S$ be the sub-family. Then $E$, the envelope of the sub-family, will touch each member of the sub-family. As $C$ lies entirely on $E$, it will touch each member of the sub-family for some $s$. This requires the relation

$$F(x_0(s), y_0(s), z_0(s), a, b) = 0, \tag{1.9.3}$$

to be satisfied for each $a$ and $b$ belonging to the corresponding sub-family, for some $s$ (this is the condition for $C$ to intersect a member of the sub-family) and also the relation

$$\frac{\partial F}{\partial s}(x_0(s), y_0(s), z_0(s), a, b) = 0, \tag{1.9.4}$$

to be satisfied for the same $s$. (This is the condition that at this point of intersection, the curve is actually a tangent to that member.)

On eliminating $s$ between $(1.9.3)$ and $(1.9.4)$, we get the required relation between $a$ and $b$. This defines the sub-family we are looking for. However there could be many solutions. Since on eliminating $s$ we get

$$\psi(a, b) = 0, \tag{1.9.5}$$

Equation (1.9.5) may be factored into a set of alternative equations $b = \phi_1(a), b = \phi_2(a)$, etc. Each one of them defines a sub-family and the envelope of each of these sub-families, if it exists, is a solution of the problem. Hence a solution, if it exists, may not be unique.

Observe that this does not happen in quasi-linear equations, if the curve $C$ is properly chosen.

**Example 1.9.3**: Find a complete integral of the equation

$$(p^2 + q^2)x = pz,$$

and the integral surface containing the curve $C \colon x_0 = 0, y_0 = s^2, z_0 = 2s$.

**Solution**: The auxiliary equations arc

$$\frac{dx}{2px - z} = \frac{dy}{2qx} = \frac{dz}{pz} = \frac{dp}{-q^2} = \frac{dq}{pq},$$

$$\frac{dp}{-q^2} = \frac{dq}{pq} \Rightarrow p^2 + q^2 = a^2.$$

Therefore

$$p = \frac{a^2 x}{z}, \quad q = \pm \frac{a\sqrt{z^2 - a^2 x^2}}{z},$$

$$dz = \frac{a^2 x}{z} dx \pm \frac{a\sqrt{z^2 - a^2 x^2}}{z} dy,$$

which implies

$$\frac{z\,dz - a^2 x\,dx}{\sqrt{z^2 - a^2 x^2}} = \pm a\,dy.$$

On integrating, the complete integral is found to be

$$z^2 = a^2 x^2 + (ay + b)^2. \tag{1.9.6}$$

Equation (1.9.3) in this case becomes

$$4s^2 = (as^2 + b)^2. \tag{1.9.7}$$

Then on differentiating (1.9.7) with respect to $s$, we get

$$2 = a(as^2 + b). \tag{1.9.8}$$

Further, on eliminating $s$ between (1.9.7) and (1.9.8), we get $ab = 1$.
Therefore, on substituting $b = 1/a$ in (1.9.6), the one-parameter sub-family of the complete integral is

$$z^2 = a^2x^2 + (ay + \frac{1}{a})^2,$$

or $\qquad\qquad a^4(x^2 + y^2) + a^2(2y - z^2) + 1 = 0, \qquad\qquad (1.9.9)$

and its envelope is obtained by eliminating $a$ between (1.9.9) and

$$2a^2(x^2 + y^2) + (2y - z^2) = 0. \qquad\qquad (1.9.10)$$

The envelope is $(2y - z^2)^2 = 4(x^2 + y^2)$, i.e., $z^2 = 2(y \pm \sqrt{x^2 + y^2})$.
Since $\sqrt{x^2 + y^2} \geq y$ , the minus sign is to be discarded and then

$$z^2 = 2(y + \sqrt{x^2 + y^2}),$$

which is the required integral surface. $\qquad\qquad\qquad\qquad\qquad\qquad$ □

**Example 1.9.4**: Find the complete integral of the equation

$$p^2x + qy - z = 0,$$

and derive the equation of the integral surface containing the line $y = 1, x + z = 0$.
**Solution**: The auxiliary equations are

$$\frac{dx}{2px} = \frac{dy}{y} = \frac{dz}{2p^2x + qy} = \frac{dp}{-p(1+p)} = \frac{dq}{0} .$$

Therefore

$$q = a \quad \text{and} \quad p = \pm \left(\frac{z - ay}{x}\right)^{1/2} .$$

On substituting for $p$ and $q$ in $dz = pdx + qdy$, we obtain

$$\frac{dz - ady}{\sqrt{z - ay}} = \pm\frac{dx}{\sqrt{x}} .$$

On integrating, we obtain

$$\sqrt{z - ay} = \pm\sqrt{x} + \sqrt{b} ,$$

i.e., $\qquad\qquad (ay - z + x + b)^2 = 4bx, \qquad\qquad (1.9.11)$

as the complete integral.

The parametric form of the given line is $x = s, y = 1, z = -s$. Equation (1.9.3) in this case takes the following form

$$(a + b + 2s)^2 = 4bs.$$

Then, on differentiating with respect to $s$ , we get

$$4(a + b + 2s) = 4b.$$

Eliminating $s$ between them gives $b^2 + 2ab = 0$, i.e., $b = 0$ or $b = -2a$. The case $b = 0$ does not lead to a solution.
On substituting $b = -2a$ in (1.9.11), we get the sub-family

$$(ay - z + x - 2a)^2 = -8ax. \tag{1.9.12}$$

The envelope of (1.9.12) is obtained on eliminating $a$ between (1.9.12) and

$$(y - 2)(ay - z + x - 2a) = -4x. \tag{1.9.13}$$

The envelope of (1.9.12) is
$$xy = z(y - 2),$$

and this is the required integral surface. □
We now show that we can derive any other complete integral from a given one in a similar way. Let
$$F(x, y, z, a, b) = 0, \tag{1.9.14}$$

be a complete integral of $f(x, y, z, p, q) = 0$. Suppose

$$G(x, y, z, h, k) = 0, \tag{1.9.15}$$

be any other complete integral of the same p.d.e. involving two arbitrary constants $h$ and $k$. We will now show that it is possible to derive (1.9.15) as an envelope of a one-parameter sub-family of (1.9.14). We first choose a curve $C$ on the surface (1.9.15) such that the constants $h$ and $k$ are independent parameters in its equations. Then we find the envelope of a one-parameter sub-family of (1.9.14) touching the curve $C$. This solution containing two arbitrary constants $h$ and $k$, is thus a complete integral.

**Example 1.9.5:**  Show that the differential equation

$$2xz + q^2 = x(xp + yq),$$

has a complete integral
$$z + a^2x = axy + bx^2, \tag{1.9.16}$$

and deduce that $x(y + hx)^2 = 4(z - kx^2)$ is also a complete integral.

**Solution**: It can be verified that Equation (1.9.16) satisfies the given differential equation. The corresponding matrix (1.2.5)

$$\begin{pmatrix} xy - 2ax & y - 2a & x \\ x^2 & 2x & 0 \end{pmatrix}$$

is of rank two if $x \neq 0$. Hence (1.9.16) is a complete integral in any domain that does not contain the line $x = 0$. Let us now take a curve on the surface $x(y + hx)^2 = 4(z - kx^2)$ involving $h$ and $k$ as

$$x_0 = s, \; y_0 = -hs, \; z_0 = ks^2.$$

Therefore we want to find a sub-family of (1.9.16) that touches this curve. Hence

$$ks^2 + a^2 s = as(-hs) + bs^2,$$

i.e.,
$$ks + a^2 = -ahs + bs.$$

On differentiating with respect to $s$, we get $k = -ah + b$. Therefore $b = k + ah$. Hence the sub-family of (1.9.16) is

$$z + a^2 x = axy + (k + ah)x^2. \tag{1.9.17}$$

To find the envelope of (1.9.17) we have to eliminate $a$ between (1.9.17) and

$$2ax = xy + hx^2. \tag{1.9.18}$$

Equation (1.9.18) gives $a = (y + hx)/2$. On substituting for $a$ in (1.9.17), we get the required envelope as
$$4(z - kx^2) = x(y + hx)^2,$$

which is the required complete integral.    □

**Exercise 1.9.1**: Find the integral surface of the differential equation $(x - y)p + (y - x - z)q = z$, passing through the circle $z = 1$, $x^2 + y^2 = 1$.

**Exercise 1.9.2**: Find the integral surface of the differential equation $x(z + 2)p + (xz + 2yz + 2y)q = z(z + 1)$, passing through the curve $x_0 = s^2$, $y_0 = 0$, and $z_0 = 2s$.

**Exercise 1.9.3**: Find the general integral of the differential equation $(x - y)y^2 p + (y - x)x^2 q = (x^2 + y^2)z$ and the particular solution through $xz = a^2$, $y = 0$.

**Exercise 1.9.4**: Find the integral surface of the differential equation $p^2 x + pqy = 2pz + x$, passing through the line $y = 1$, $x = z$.

**Exercise 1.9.5**: Find a solution of $p + q^2 = 0$, passing through the line $x_0(s) = 0, y_0(s) = s, z_0(s) = 3s$.

**Exercise 1.9.6**: Find a solution of $z = p^2 - q^2$, which passes through the curve $C$ with the equation $4z + x^2 = 0, y = 0$.

**Exercise 1.9.7**: Find the integral surface of the p.d.e. $pq = z$ containing the curve $x = 0, z = y^2$.

# 1.10  Quasi-Linear Equations: Geometry of Solutions

We shall first discuss the semi-linear case.

**Semi-linear equations**:

Consider a semi-linear equation

$$P(x, y)z_x + Q(x, y)z_y = R(x, y, z), \tag{1.10.1}$$

where $P, Q$, and $R$ are continuously differentiable functions and both $P(x, y)$ and $Q(x, y)$ do not vanish simultaneously. Note that the expression on the left-hand side of (1.10.1) is the directional derivative of $z(x, y)$ in the direction $(P(x, y), Q(x, y))$ at the point $(x, y)$.

Let us now consider the one-parameter family of curves in the $x, y$-plane defined by the ordinary differential equation

$$\frac{dy}{dx} = \frac{Q(x, y)}{P(x, y)},$$

or the system of ordinary differential equations

$$\frac{dx}{dt} = P(x, y), \frac{dy}{dt} = Q(x, y). \tag{1.10.2}$$

These curves have the property that along them $z(x, y)$ will satisfy the ordinary differential equation

$$\frac{dz}{dx} = z_x + z_y \frac{dy}{dx} = \frac{P(x, y)z_x + Q(x, y)z_y}{P(x, y)},$$

or

$$\frac{dz}{dt} = z_x \frac{dx}{dt} + z_y \frac{dy}{dt} = z_x P(x, y) + z_y Q(x, y) = R(x, y, z). \tag{1.10.3}$$

The one-parameter family of curves $C_{\lambda_0}$ defined by Equations (1.10.2) are called the characteristic curves of the partial differential equation (1.10.1).

Let $(x_0, y_0)$ be a point in the $x, y$-plane. By the existence and uniqueness of the solution of the initial value problem for the ordinary differential equations, Equation (1.10.2) will define a unique characteristic curve

$$x(t) = x(x_0, y_0, t), \quad y(t) = y(x_0, y_0, t), \tag{1.10.4}$$

such that $x(0) = x_0$ and $y(0) = y_0$. Suppose we now assign a value $z_0$ for $z(x, y)$ at $(x_0, y_0)$. Then the Equation (1.10.4) determines a unique solution $z$ as

$$z = z(x_0, y_0, t), \tag{1.10.5}$$

such that $z(x_0, y_0, 0) = z_0$. That is, $z(x, y)$ is uniquely determined along the whole characteristic passing through $(x_0, y_0)$ if we assign a value for $z$ at $(x_0, y_0)$.

Hence if $z(x, y)$ is known on a curve $\Gamma_0$ in the $x, y$-plane and if this curve is such that it intersects the one-parameter family of characteristic curves $C_{\lambda_0}$, we can determine $z(x, y)$ uniquely in the region covered by $C_{\lambda_0}$.

Observe that the curve $\Gamma_0$ cannot be chosen arbitrarily. We will deal with this point later (refer to Theorem 1.10.1).

**Example 1.10.1**: Solve $xz_y - yz_x = z$ with the initial condition $z(x, 0) = f(x)$, $x \geq 0$.
**Solution**: The characteristic curves are given by the equation

$$\frac{dy}{dx} = -\frac{x}{y},$$

having the solution $x^2 + y^2 = c^2$. Along such a curve, $z$ satisfies the ordinary differential equation

$$\frac{dz}{dx} = -\frac{z}{y} = -\frac{z}{\sqrt{c^2 - x^2}},$$

i.e.,

$$\frac{dz}{z} = -\frac{dx}{\sqrt{c^2 - x^2}},$$

whose solution is

$$z = k(c)e^{-\sin^{-1}(x/c)},$$

where the arbitrary function $k$ may depend on $c$. Hence we have the general solution

$$z = k(x^2 + y^2)e^{-\sin^{-1}(x/c)}.$$

On applying the initial conditions, we get

$$f(x) = k(x^2)e^{-\pi/2},$$
$$k(x) = f(\sqrt{x})e^{\pi/2}.$$

Hence the required solution is

$$z(x, y) = f(\sqrt{x^2 + y^2})e^{\pi/2 - \sin^{-1}(x/c)}. \qquad \square$$

**Quasi-linear equations**: We now consider the quasi-linear equation

$$P(x, y, z)z_x + Q(x, y, z)z_y = R(x, y, z), \qquad (1.10.6)$$

where $P, Q$, and $R$ are continuously differentiable functions of $x, y$, and $z$ and $P, Q$, and $R$ do not vanish simultaneously. Its solution $z(x, y)$ defines an integral surface $z = z(x, y)$ in the $x, y, z$-space. The normal to this surface has direction ratios $(z_x, z_y, -1)$. Hence Equation (1.10.6) states the condition that the integral surface is such that at each point, the line with direction ratios $(P, Q, R)$ is tangent to the surface at that point. In fact, any surface $z = z(x, y)$ is an integral surface if and only if the tangent plane contains the characteristic direction $(P, Q, R)$, which is a direction field defined by the p.d.e. at each point.

We consider next the integral curves of this field called the characteristic curves. They are a family of space curves whose tangent at each point coincides with the characteristic direction $(P, Q, R)$ at that point and are given by the following system of ordinary differential equations

$$\frac{dx}{P(x, y, z)} = \frac{dy}{Q(x, y, z)} = \frac{dz}{R(x, y, z)} = dt \text{ (say)}, \qquad (1.10.7)$$

or

$$\left.\begin{array}{l} \dfrac{dx}{dt} = P(x, y, z), \\ \dfrac{dy}{dt} = Q(x, y, z), \\ \dfrac{dz}{dt} = R(x, y, z). \end{array}\right\} \qquad (1.10.8)$$

By the existence and uniqueness of the solution of the initial value problem for a system of ordinary differential equations, there passes a characteristic curve $x = x(x_0, y_0, z_0, t), y = y(x_0, y_0, z_0, t), z = z(x_0, y_0, z_0, t)$ through each point $(x_0, y_0, z_0)$.

The system of ordinary differential equations (1.10.8) being an autonomous system, there will be a two-parameter family of integral curves.

Hence there is a two-parameter family of characteristic curves. Every surface generated by a one-parameter family of characteristics is an integral surface. For example, if we consider any point on such a surface, then the tangent to the characteristic curve passing through that point lies on the tangent plane to the surface. Thus the tangent plane to the surface at each point contains the line with direction ratios $(P, Q, R)$. Hence the surface is an integral surface.

In addition, the converse is also true, i.e., every integral surface is generated by a family of characteristic curves. For, let us consider an integral surface $S$ given by $z = z(x, y)$. Consider any point $(x_0, y_0, z_0)$ on $S$. Let the solution of

$$\frac{dx}{dt} = P(x, y, z(x, y)), \frac{dy}{dt} = Q(x, y, z(x, y)),$$

with the initial conditions $x = x_0, y = y_0$ at $t = 0$ be given by $x = x(t)$ and $y = y(t)$. Consider the corresponding curve in three dimensions

$$x = x(t), \quad y = y(t), \quad z = z(x(t), y(t)).$$

Observe that this curve lies on the given integral surface. Moreover,

$$\frac{dz}{dt} = z_x \frac{dx}{dt} + z_y \frac{dy}{dt} = P(x, y, z)z_x + Q(x, y, z)z_y,$$
$$= R(x, y, z).$$

Therefore the curve satisfies Equation (1.10.8) for characteristic curves and it passes through the point $(x_0, y_0, z_0)$. Therefore $S$ is generated by the characteristic curves as it contains the characteristic curve through each point $(x_0, y_0, z_0)$ on $S$. Further, if a characteristic curve has one point in common with the integral surface, it lies entirely on the integral surface since only one characteristic curve passes through a given point. In addition, if two integral surfaces intersect at a point, then they intersect along the entire characteristic curve through this point. Hence the curve of intersection of two integral surfaces must be a characteristic curve.

**Note**: The two-parameter family of characteristics is nothing but the curves of intersection of the surfaces $u = c_1$ and $v = c_2$ given by (1.4.3). To find an integral surface passing through a given curve, as in Section 1.9, we find a relation between $c_1$ and $c_2$, which means we are actually choosing a one-parameter sub-family of characteristics, which in turn generates the required integral surface.

**Theorem 1.10.1**: Consider the first order quasi-linear partial differential equation

$$P(x, y, z)z_x + Q(x, y, z)z_y = R(x, y, z), \qquad (1.10.9)$$

where $P, Q$, and $R$ have continuous partial derivatives with respect to $x, y$, and $z$ and they do not vanish simultaneously. Let the value $z = z_0(s)$ be prescribed along the initial curve $\Gamma_0 : x = x_0(s), y = y_0(s)$, $x_0, y_0$, and $z_0$ being continuously differentiable functions $(a \leq s \leq b)$. Further, for $a \leq s \leq b$, if

$$\frac{dy_0}{ds}P(x_0(s), y_0(s), z_0(s)) - \frac{dx_0}{ds}Q(x_0(s), y_0(s), z_0(s)) \neq 0, \qquad (1.10.10)$$

then there exists a unique solution $z(x, y)$ defined in some neighborhood of the initial curve $\Gamma_0$, which satisfies the p.d.e. and the initial condition

$$z(x_0(s), y_0(s)) = z_0(s),$$

i.e., the integral surface $z = z(x, y)$ contains the initial data curve $C : x = x_0(s), y = y_0(s), z = z_0(s)$.

**Note**: The initial curve $\Gamma_0$ is the projection of the initial data curve $C$ on the $x, y$-plane.

**Proof**: The ordinary differential equations (1.10.8) can be solved to obtain a unique family of characteristics,

$$\left.\begin{array}{l} x = x(x_0, y_0, z_0, t) = x(s, t), \\ y = y(x_0, y_0, z_0, t) = y(s, t), \\ z = z(x_0, y_0, z_0, t) = z(s, t), \end{array}\right\} \qquad (1.10.11)$$

having continuous derivatives with respect to the parameters $s$ and $t$, satisfying the initial conditions $x(s, 0) = x_0(s), y(s, 0) = y_0(s)$, and $z(s, 0) = z_0(s)$ (from the existence and uniqueness theorem for the ordinary differential equations).

Observe that the Jacobian

$$\frac{\partial(x, y)}{\partial(s, t)}\bigg|_{t=0} = \begin{vmatrix} x_s & x_t \\ y_s & y_t \end{vmatrix}_{t=0} = (x_s Q - y_s P)_{t=0} \neq 0,$$

from (1.10.10). Hence we can solve the first two equations of (1.10.11) for $s$ and $t$ in terms of $x$ and $y$ in a neighborhood of the initial curve $t = 0$, i.e., $s = s(x, y), t = t(x, y)$.

Let $z = \phi(x, y) = z(s(x, y), t(x, y))$. Note that, $\phi(x, y)$ satisfies the initial condition as $\phi(x_0, y_0) \equiv z(s, 0) = z_0(s)$. Moreover, $z = \phi(x, y)$ satisfies the p.d.e. (1.10.9).

For, consider

$$
\begin{aligned}
P\phi_x + Q\phi_y &= P(z_s s_x + z_t t_x) + Q(z_s s_y + z_t t_y),\\
&= z_s(Ps_x + Qs_y) + z_t(Pt_x + Qt_y),\\
&= z_s(s_x x_t + s_y y_t) + z_t(t_x x_t + t_y y_t).
\end{aligned}
$$

Observe that

$$
s_x x_t + s_y y_t = s_t = 0,
$$

$$
t_x x_t + t_y y_t = t_t = 1.
$$

Hence

$$
P\phi_x + Q\phi_y = z_t = R.
$$

The uniqueness of $\phi(x, y)$ follows from the following argument. Suppose $\phi(x, y)$ is not unique, then there are two integral surfaces that intersect along the given initial data curve $C$. Then through each point on the initial data curve, there passes one and only one characteristic curve. Therefore this characteristic curve has to lie on both the surfaces. Hence the same family of characteristic curves that pass through each point of the initial data curve lie on both the surfaces. Hence both the surfaces must coincide as both are generated by the same family of characteristic curves. Thus $\phi(x, y)$ is unique. $\qquad\square$

**Note**: The condition (1.10.10) is said to be the admissibility condition and any initial data curve $C$ satisfying this condition is said to be an admissible curve.

Hence the integral surface (the solution) is generated by the one parameter family of characteristics $C_\lambda$ issuing from each point of the initial data curve $C$ (refer to Fig. 1.10.1).

**Example 1.10.2**: Solve the initial value problem for the quasi-linear equation $zz_x + z_y = 1$ containing the initial data curve $C : x_0 = s,\ y_0 = s,\ z_0 = \frac{1}{2}s$ for $0 \le s \le 1$.

**Solution**: Observe that

$$
\frac{dy_0}{ds}P - \frac{dx_0}{ds}Q = \frac{1}{2}s - 1 \neq 0 \quad \text{for } 0 \le s \le 1.
$$

Solving the following system of ordinary differential equations

$$
\frac{dx}{dt} = z,\ \frac{dy}{dt} = 1,\ \frac{dz}{dt} = 1,
$$

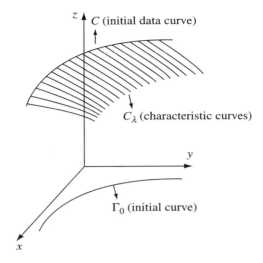

Figure 1.10.1

with initial conditions $x(s,0) = s$, $y(s,0) = s$, $z(s,0) = \frac{1}{2}s$, the family of characteristics through the initial data curve is found to be

$$x = \frac{1}{2}t^2 + \frac{1}{2}st + s,$$
$$y = t + s,$$
$$z = t + \frac{s}{2}.$$

Solving for $s$ and $t$ in terms of $x$ and $y$, we obtain

$$s = \frac{x - (y^2/2)}{1 - (y/2)}, \quad t = \frac{y - x}{1 - (y/2)}.$$

Hence the solution is

$$z = \frac{2(y-x) + (x - y^2/2)}{2 - y} = \frac{2y - x - \frac{y}{2}}{2 - y} = \frac{4y - 2x - y^2}{2(2 - y)}. \qquad \square$$

**Example 1.10.3**: Find the integral surface of $zz_x + z_y = 0$ containing the initial data curve $C : x_0 = s, y_0 = 0, z_0 = f(s)$, where

$$f(s) = \begin{cases} 1, & s \le 0, \\ 1 - s, & 0 \le s \le 1, \\ 0, & s \ge 1. \end{cases}$$

Here we observe that

$$\frac{dy_0}{ds}P - \frac{dx_0}{ds}Q = -1 \neq 0 \quad \forall \quad s.$$

Solving the following system of ordinary differential equations

$$\frac{dx}{dt} = z, \frac{dy}{dt} = 1, \frac{dz}{dt} = 0,$$

with the initial conditions $x(s,0) = s$, $y(s,0) = 0$, $z(s,0) = f(s)$, the family of characteristics through the initial data curve is found to be $x = tf(s) + s, y = t, z = f(s)$. (Refer to Example 1.4.4 as well.) Hence

$$x(s,t) = \begin{cases} s+t, & s \leq 0, \\ (1-s)t + s, & 0 \leq s \leq 1, \\ s, & s \geq 1, \end{cases} \quad z(s,t) = \begin{cases} 1, & s \leq 0, \\ 1-s, & 0 \leq s \leq 1, \\ 0, & s \geq 1. \end{cases}$$

Here $x = yf(s)+s$. Hence the characteristics are straight lines intersecting the $x$-axis at $(s,0)$. The slope of the characteristic through $(s,0)$ is $\frac{1}{f(s)}$ and $z$ is a constant on each of them. We observe that in the interval $0 \leq s \leq 1$, $f(s)$ is a decreasing function of $s$. So the slopes of these characteristics increase with $s$ and hence they intersect in this region.

On a characteristic issuing from a point $(s,0)$, $z$ takes the value $(1-s)$ in $0 \leq s \leq 1$. Thus $z$ takes a different value on each characteristic and as all these lines intersect at $(1,1)$, $z$ is multi-valued at $(1,1)$ (refer to Fig. 1.10.2). Thus the solution cannot be single valued and breaks down at $(1,1)$, i.e., it cannot be defined uniquely.

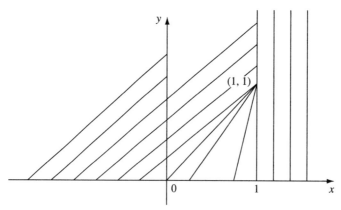

Figure 1.10.2

The characteristics issuing from $(s, 0)$ with $s < 0$ intersect those issuing from $(s, 0)$ with $s > 1$ and since $z$ has different values on them, the solution is not defined in the quadrant $x \geq 1, y \geq 1$. $\qquad\square$

**Example 1.10.4**: Solve the Cauchy problem for $2z_x + yz_y = z$ for the initial data curve $C : x_0 = s, y_0 = s^2, z_0 = s, 1 \leq s \leq 2$.

**Solution**: We observe that

$$\frac{dy_0}{ds}P - \frac{dx_0}{ds}Q = 4s - s^2 \neq 0,$$

for $1 \leq s \leq 2$.

Therefore $C$ is admissible for $1 \leq s \leq 2$. We solve

$$\frac{dx}{dt} = 2, \quad \frac{dy}{dt} = y, \quad \frac{dz}{dt} = z,$$

such that $x(s, 0) = s, y(s, 0) = s^2, z(s, 0) = s$.

The solutions of the system of ordinary differential equations that determine the characteristics through the initial data curve are

$$x = s + 2t, \quad y = s^2 e^t, \quad z = se^t.$$

The solution is obtained by eliminating $s$ and $t$, i.e.,

$$z^2 = y \exp[(xz - y)/2z]. \qquad\square$$

**Example 1.10.5**: Find the solution of the initial value problem for the quasi-linear equation $z_x - zz_y + z = 0$ for all $y$ and $x > 0$, for the initial data curve $C : x_0 = 0, y_0 = s, z_0 = -2s, -\infty < s < \infty$.

**Solution**: The admissibility condition is satisfied since

$$\frac{dy_0}{ds}P - \frac{dx_0}{ds}Q = 1 \neq 0 \ \forall \ s.$$

The characteristic curves that generate the surface are obtained from the solutions of

$$\frac{dx}{dt} = 1, \quad \frac{dy}{dt} = -z, \quad \frac{dz}{dt} = -z,$$

satisfying the following conditions

$$x(s, 0) = 0, \quad y(s, 0) = s, \quad z(s, 0) = -2s.$$

The solutions are found to be

$$x = t, \quad y = -2se^{-t} + 3s, \quad z = -2se^{-t}.$$

The parameters $s$ and $t$ in terms of $x$ and $y$ are found to be

$$s = -\frac{y}{2e^{-x} - 3}, \quad t = x.$$

Therefore

$$z = \frac{2y}{2e^{-x} - 3}e^{-x},$$

$$= -\frac{2y}{3e^x - 2}, \quad \log(2/3) > x \geq 0.$$

The solution breaks down at $x = \log(2/3)$.  □

**Note**: In the case of quasi-linear equations, the solution of the Cauchy problem is unique if the initial curve is smooth and is not parallel to the characteristic curve (i.e., it satisfies the condition (1.10.10)). However, the same is not true in the case of non-linear equations as seen in the following example.

**Example 1.10.6**: Find the solution of the equation

$$z = \frac{1}{2}(z_x^2 + z_y^2) + (z_x - x)(z_y - y),$$

which passes through the $x$ - axis.

**Ans. 1**: $z = y(4x - 3y)/2$.

**Ans. 2**: $z = y^2/2$.

Observe that the previous two surfaces pass through the $x$- axis and satisfy the given equation. Hence this problem has no unique solution. Refer to Example (1.11.2). □

Non-linear equations, unlike quasi-linear equations may have many solutions or no solution for the Cauchy problem. One needs some more information to have a unique solution for the Cauchy problem in the non-linear case. We discuss this problem in the next section.

**Exercise 1.10.1**: Find the integral surface passing through the initial data curve $C : x_0 = -1, y_0 = s, z_0 = \sqrt{s}$ of the equation

$$(x + 2)p + 2yq = 2z.$$

**Ans.**: $z = \sqrt{y}(x + 2)$.

**Exercise 1.10.2**: Solve $z_x + z_y = z^2$ with the initial condition $z(x, 0) = f(x)$.

**Ans.**: $z = \dfrac{f(x - y)}{(1 - yf(x - y))}$.

**Exercise 1.10.3**: Find the integral surface passing through the initial data curve $x = 1$, $z = y^2 + y$ of the equation
$$x^3 z_x + y(3x^2 + y)z_y = z(2x^2 + y).$$
**Ans.**: $(x^2 + y)(xz - y) = yz$.

**Exercise 1.10.4**: Find the integral surface for the differential equation $z(xz_x - yz_y) = y^2 - x^2$ passing through the initial data curve $(2s, s, s)$.
**Ans.**: $z^2 = 3xy - x^2 - y^2$.

# 1.11   Non-linear First Order P.D.E.

Consider a non-linear first order p.d.e.

$$f(x, y, z, p, q) = 0. \tag{1.11.1}$$

We assume that $f$ has continuous second order derivatives with respect to its variables $x, y, z, p$, and $q$ and either $f_p$ or $f_q$ is non-zero at every point. Without loss of generality, let us assume that $f_q \neq 0$, so that we can solve (1.11.1) at each point $(x, y, z)$ for $q$ as $q = q(x, y, z, p)$. Of course, $q$ need not always be a single valued function of $p$, but it may be assumed that one branch of a possible set of solutions $q = q(x, y, z, p)$ is chosen.

**Monge Cone:**   Let $(x_0, y_0, z_0)$ be some point in space. We consider a possible integral surface $z = z(x, y)$ and the direction ratios $(p, q, -1)$ of the normal to the tangent plane
$$z - z_0 = p(x - x_0) + q(y - y_0), \tag{1.11.2}$$
to the integral surface at that point. The equation (1.11.1), which is equivalent to

$$q = q(x_0, y_0, z_0, p), \tag{1.11.3}$$

indicates that $p$ and $q$ are not independent at $(x_0, y_0, z_0)$. This implies that the integral surfaces are those surfaces having tangent planes belonging to a one-parameter $(p)$ family.

Consider the family of planes given by Equation (1.11.2) passing through $(x_0, y_0, z_0)$ as $p$ varies and $q$ is determined by Equation (1.11.3). These planes envelope a cone called the Monge cone having its vertex at $(x_0, y_0, z_0)$. Thus if $z = z(x, y)$ is a solution of (1.11.1) it must be tangent to the Monge cone at each point $(x, y, z)$ on

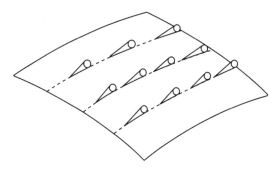

Figure 1.11.1

the surface. Thus the differential equation (1.11.1) characterizes a field of cones such that a surface will be an integral surface if and only if it is tangent to the Monge cone at each point (refer to Fig. 1.11.1). The line of contact of the Monge cone with the tangent plane to the integral surface at each point gives a field of directions. These directions are called the characteristic directions at that point and lie along the generators of the Monge cone. The integral curves of this field of directions on the surface define a family of curves called the characteristic curves.

**Analytic expression for the Monge cone at $(x_0, y_0, z_0)$:**
    As the Monge cone at $(x_0, y_0, z_0)$ is the envelope of the one-parameter family of planes

$$z - z_0 = p(x - x_0) + q(y - y_0),$$

where $q$ is given by

$$q = q(x_0, y_0, z_0, p), \tag{1.11.3}$$

it can be obtained by eliminating $p$ from the following equations

$$z - z_0 = p(x - x_0) + q(x_0, y_0, z_0, p)(y - y_0), \tag{1.11.4}$$

and
$$0 = (x - x_0) + (y - y_0)\frac{dq}{dp}. \tag{1.11.5}$$

On differentiating (1.11.1) with respect to $p$, we get

$$\frac{df}{dp} = f_p + f_q\frac{dq}{dp} = 0,$$

as $q$ is a function of $p$. If one eliminates $\dfrac{dq}{dp}$ from (1.11.5) and the previous equation, then the equations describing the Monge cone can be written as

$$\left.\begin{array}{rcl} q & = & q(x_0, y_0, z_0, p), \\[2mm] z - z_0 & = & p(x - x_0) + q(y - y_0), \\[2mm] \dfrac{x - x_0}{f_p} & = & \dfrac{y - y_0}{f_q}. \end{array}\right\} \tag{1.11.6}$$

The last two equations define a generator of the cone for a given $p$ and the corresponding $q$ given by Equation (1.11.3), i.e., the line of contact between the tangent plane and the cone.

**Example 1.11.1**: Consider $p^2 + q^2 = 1$. Let $(x_0, y_0, z_0) = (0, 0, 0)$. Then the Monge cone is obtained by eliminating $p$ and $q$ from

$$q = \sqrt{1 - p^2},$$

$$z = px + qy,$$

$$\frac{x}{2p} = \frac{y}{2q},$$

which is $x^2 + y^2 = z^2$. This is a cone with vertex at $(0, 0, 0)$ (as every second degree homogeneous equation in $x, y$, and $z$ represents a cone with vertex at $(0, 0, 0)$).   □

**Note**: Observe that in the quasi-linear case $Pp + Qq = R$, every tangent plane (1.11.2) at $(x_0, y_0, z_0)$ contains the following line (why?)

$$\frac{x - x_0}{P} = \frac{y - y_0}{Q} = \frac{z - z_0}{R}.$$

Hence the Monge cone degenerates into a straight line given by the previous equation at each point.

Let us now find the equations that give the characteristic curves. As the line of contact between the tangent plane to an integral surface and the Monge cone at that point is the generator of the Monge cone, it can be written from (1.11.6) as

$$\frac{x - x_0}{f_p} = \frac{y - y_0}{f_q} = \frac{z - z_0}{pf_p + qf_q}. \tag{1.11.7}$$

Hence the characteristic directions that lie along the generators of the Monge cone are given by $(f_p, f_q, pf_p + qf_q)$.

The characteristic curves are thus the solutions of

$$\frac{dx}{f_p} = \frac{dy}{f_q} = \frac{dz}{pf_p + qf_q},$$

or

$$\frac{dx}{dt} = f_p \,, \quad \frac{dy}{dt} = f_q \,, \quad \frac{dz}{dt} = pf_p + qf_q. \tag{1.11.8}$$

Since these three ordinary differential equations for determining the characteristic curve $(x(t), y(t), z(t))$ cannot be solved because they involve the functions $p$ and $q$, we require more information regarding the variation of $p$ and $q$ along a characteristic curve. Along such a curve on the given integral surface

$$\left. \begin{aligned} \frac{dp}{dt} &= p_x \frac{dx}{dt} + p_y \frac{dy}{dt} = p_x f_p + p_y f_q, \\ \frac{dq}{dt} &= q_x \frac{dx}{dt} + q_y \frac{dy}{dt} = q_x f_p + q_y f_q. \end{aligned} \right\} \tag{1.11.9}$$

Further, by differentiating Equation (1.11.1) with respect to $x$ as well as $y$, we get respectively

$$f_x + f_z p + f_p p_x + f_q q_x = 0,$$

$$f_y + f_z q + f_p p_y + f_q q_y = 0.$$

Since $p_y = q_x$, Equations (1.11.9) may now be rewritten as

$$\frac{dp}{dt} = -f_x - f_z p, \frac{dq}{dt} = -f_y - f_z q. \tag{1.11.10}$$

Thus for a given integral surface $z = z(x, y)$, there corresponds a family of characteristic curves on the surfaces associated with it such that the coordinates of the curve $(x(t), y(t), z(t))$ and the numbers $p(t)$ and $q(t)$ along the curve are related by a system of five ordinary differential equations given by (1.11.8) and (1.11.10). These five ordinary differential equations are called the characteristic differential equations related to the given p.d.e. (1.11.1).

In order to determine the integral surface, by the previous discussion, we consider the p.d.e. (1.11.1) along with the system of characteristic equations (1.11.8) and

(1.11.10) as a system of six equations, i.e.,

$$
\left.
\begin{aligned}
&f(x, y, z, p, q) = 0 \ , \\
&\frac{dx}{dt} = f_p \ , \\
&\frac{dy}{dt} = f_q \ , \\
&\frac{dz}{dt} = p f_p + q f_q \ , \\
&\frac{dp}{dt} = -f_x - f_z p \ , \\
&\frac{dq}{dt} = -f_y - f_z q \ ,
\end{aligned}
\right\}
\tag{1.11.11}
$$

for the five unknown functions $x(t), y(t), z(t), p(t)$, and $q(t)$. This system is not over-determined. Consider a solution of the last five equations of (1.11.11). Along such a solution we find that

$$
\begin{aligned}
\frac{df}{dt} &= f_x \frac{dx}{dt} + f_y \frac{dy}{dt} + f_z \frac{dz}{dt} + f_p \frac{dp}{dt} + f_q \frac{dq}{dt} \ , \\
&= f_x f_p + f_y f_q + p f_z f_p + q f_z f_q - f_p f_x - f_p f_z p - f_q f_y - f_q f_z q \ , \\
&= 0.
\end{aligned}
$$

Hence $f$ = constant, is an integral of the system of ordinary differential equations. Thus we can conclude that the equation $f(x, y, z, p, q) = 0$ is not much of a restriction. Suppose that $f = 0$ at $(x_0, y_0, z_0, p_0, q_0)$ at $t = 0$. Then the unique solution of the characteristic equations passing through this point is such that $f = 0$ will be satisfied for all $t$ along the solution.

**Characteristic strip**:
A solution $(x(t), y(t), z(t), p(t), q(t))$ of the system of ordinary differential equations (1.11.11) can be interpreted as a strip.
The first three functions, namely, $(x(t), y(t), z(t))$ determine a space curve. At each point of the space curve, $p(t)$ and $q(t)$ define a tangent plane with $(p, q, -1)$ as the normal vector. The curve along with these tangent planes at each point is called a characteristic strip (refer to Fig. 1.11.2) and the curve is called the characteristic curve.

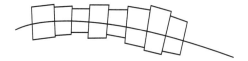

Figure 1.11.2

For a fixed $t$ given by $t_0$, $(x_0, y_0, z_0, p_0, q_0)$ is said to define an element of the strip, i.e., the point $(x_0, y_0, z_0)$ on the curve and the corresponding tangent plane whose normal is given by $(p_0, q_0, -1)$ at that point.

**Note**: Any set of five functions $x(t), y(t), z(t), p(t)$, and $q(t)$ can be interpreted as a strip only if the following condition called the 'strip condition'

$$\frac{dz}{dt}(t) = p(t)\frac{dx}{dt}(t) + q(t)\frac{dy}{dt}(t), \tag{1.11.12}$$

is satisfied. This is the condition that at each point on the space curve $x = x(t), y = y(t), z = z(t)$, the tangent to the space curve lies on the plane whose normal has direction ratios $(p(t), q(t), -1)$.

Any solution $(x(t), y(t), z(t), p(t), q(t))$ of (1.11.11) automatically satisfies the condition (1.11.12) by virtue of the second, third, and fourth equations of (1.11.11).

**Lemma 1.11.1**:

If an element $(x_0, y_0, z_0, p_0, q_0)$ is common to both an integral surface $z = z(x, y)$ and a characteristic strip, then the corresponding characteristic curve lies completely on the surface.

**Proof**: Let $z = z(x, y)$ be an integral surface. Consider the two ordinary differential equations

$$\frac{dx}{dt}(t) = f_p(x, y, z(x, y), z_x(x, y), z_y(x, y)),$$

$$\frac{dy}{dt}(t) = f_q(x, y, z(x, y), z_x(x, y), z_y(x, y)),$$

for $x(t), y(t)$ with the initial conditions $x(0) = x_0, y(0) = y_0$, which uniquely determine a curve $x = x(t)$ and $y = y(t)$ in the $x, y$-plane. Consider the curve $x = x(t), y = y(t), z = z(t) = z(x(t), y(t))$ in space on the integral surface. Then on such a curve

$$\frac{dz}{dt} = z_x \frac{dx}{dt} + z_y \frac{dy}{dt} = z_x f_p + z_y f_q, \tag{1.11.13}$$

$$\frac{dz_x}{dt} = z_{xx} f_p + z_{xy} f_q, \tag{1.11.14}$$

$$\frac{dz_y}{dt} = z_{yx} f_p + z_{yy} f_q. \tag{1.11.15}$$

Further,

$$z(0) = z(x_0, y_0) = z_0,$$
$$z_x(0) = z_x(x_0, y_0) = p_0,$$
$$z_y(0) = z_y(x_0, y_0) = q_0.$$

Since $z = z(x, y)$ is an integral surface of the given p.d.e., we have

$$f(x, y, z(x, y), z_x(x, y), z_y(x, y)) = 0.$$

On differentiating the previous equation with respect to $x$ and $y$, we get

$$f_x + f_z z_x + f_{z_x} z_{xx} + f_{z_y} z_{yx} = 0,$$
$$f_y + f_z z_y + f_{z_x} z_{xy} + f_{z_y} z_{yy} = 0.$$

Equations (1.11.14) and (1.11.15) therefore reduce to

$$\frac{dz_x}{dt} = -f_x - f_z z_x, \tag{1.11.16}$$

$$\frac{dz_y}{dt} = -f_y - f_z z_y. \tag{1.11.17}$$

Therefore the five functions $x = x(t)$, $y = y(t)$, $z = z(x(t), y(t))$, $p = z_x(x(t), y(t))$, and $q = z_y(x(t), y(t))$ determine a characteristic strip as they satisfy the characteristic equations (1.11.11). In addition, these functions determine the unique characteristic strip with the initial element $x_0, y_0, z_0, p_0$, and $q_0$. However, this curve lies on the surface by definition. Hence the result.  □

We shall make use of this lemma while proving the uniqueness of the solution of the Cauchy problem for the non-linear partial differential equations.

**Initial strip**:
Let $x = x_0(s), y = y_0(s), z = z_0(s)$ be some arbitrary initial data curve. Suppose

that along this curve we can specify the functions $p_0(s)$ and $q_0(s)$ such that together with the initial data curve $(x_0(s), y_0(s), z_0(s))$, they satisfy the equation

$$f(x_0(s), y_0(s), z_0(s), p_0(s), q_0(s)) = 0, \qquad (1.11.18)$$

and the strip condition

$$\frac{dz_0}{ds} = p_0 \frac{dx_0}{ds} + q_0 \frac{dy_0}{ds}. \qquad (1.11.19)$$

Then such an initial element $x_0(s), y_0(s), z_0(s), p_0(s), q_0(s)$ is said to define an initial strip for the initial data curve $(x_0(s), y_0(s), z_0(s))$. The integral surface may be constructed with the help of the characteristic strips issuing from each point of the initial data curve by piecing together the characteristic curves to form a smooth surface.

As there can be more than one pair of functions $p_0$ and $q_0$ satisfying the two equations (1.11.18) and (1.11.19), there can be several initial strips associated with the same initial data curve. Hence there can be more than one integral surface passing through a given initial data curve.

**Note**: In the quasi-linear case, we note that both Equations (1.11.18) and (1.11.19) reduce to

$$P(x_0(s), y_0(s), z_0(s))p_0 + Q(x_0(s), y_0(s), z_0(s))q_0 = R(x_0, y_0, z_0),$$

and

$$\frac{dz_0}{ds} = p_0 \frac{dx_0}{ds} + q_0 \frac{dy_0}{ds},$$

which are linear in $p_0$ and $q_0$. Further, if $P\dfrac{dy_0}{ds} - Q\dfrac{dx_0}{ds} \neq 0$ then $p_0$ and $q_0$ are determined uniquely. (Refer to Theorem 1.10.1).

The previous discussion suggests a method of solving the Cauchy problem for the non-linear p.d.e. (1.11.1). The method involves the following steps.

**Step 1**: Given the p.d.e. (1.11.1) and an initial data curve $x = x_0(s)$, $y = y_0(s)$, $z = z_0(s)$, determine the functions $p_0(s)$, and $q_0(s)$ such that the five functions $x_0(s)$, $y_0(s)$, $z_0(s)$, $p_0(s)$, and $q_0(s)$ satisfy Equations (1.11.18) and (1.11.19). There could be several choices for $p_0(s)$, and $q_0(s)$. One may expect to find a unique solution for each such choice.

**Step 2**: Once a choice is made for $p_0(s)$ and $q_0(s)$, i.e., the initial strip is chosen, we can solve the last five equations of Equation (1.11.11) subject to the initial conditions $x = x_0(s)$, $y = y_0(s)$, $z = z_0(s)$, $p = p_0(s)$, and $q = q_0(s)$ at $t = 0$. Hence we have the characteristic strips issuing from each point of the initial data curve. The corresponding characteristic curves generate the required integral surface. We demonstrate the method in the following examples. We shall later prove the existence and uniqueness of the solution in Theorem 1.11.1 that actually justifies the previous method.

**Example 1.11.2**: Find the solution of the equation

$$z = \frac{1}{2}(p^2 + q^2) + (p - x)(q - y),$$

which passes through the $x$-axis.

**Solution**: Let us find the possible initial strips. The initial data curve is $x_0 = s$, $y_0 = 0$, $z_0 = 0$, and Equations (1.11.18) and (1.11.19) respectively are

$$0 = \frac{1}{2}(p_0^2 + q_0^2) + (p_0 - s)(q_0),$$
$$0 = p_0 \cdot 1 + q_0 \cdot 0.$$

The second equation implies that $p_0 = 0$.
On substituting in the first equation, we get

$$\frac{1}{2}q_0^2 + (-s)(q_0) = 0,$$

$$q_0 = 0 \quad \text{or} \quad q_0 = 2s.$$

Therefore there are two initial strips

$$\text{Case (i)} \quad x_0 = v, y_0 = 0, z_0 = 0, p_0 = 0, q_0 = 2v, \tag{1.11.20}$$

$$\text{Case (ii)} \quad x_0 = s, y_0 = 0, z_0 = 0, p_0 = 0, q_0 = 0. \tag{1.11.21}$$

The characteristic equations of this p.d.e. are given by

$$\frac{dx}{dt} = p + q - y,$$

$$\frac{dy}{dt} = p + q - x,$$

$$\frac{dz}{dt} = p(p + q - y) + q(p + q - x),$$

$$\frac{dp}{dt} = p + q - y,$$

$$\frac{dq}{dt} = p + q - x.$$

Therefore

$$\frac{d}{dt}(x - p) = 0 \quad \text{and} \quad \frac{d}{dt}(y - q) = 0. \tag{1.11.22}$$

In Case (i), Equations (1.11.22) give $x = v + p, \quad y = q - 2v$.
Observe that

$$\frac{d}{dt}(p + q - x) = p + q - x,$$

$$p + q - x = ve^t.$$

Similarly $p + q - y = 2ve^t$. Hence

$$x = v(2e^t - 1), \quad y = v(e^t - 1), \tag{1.11.23}$$

$$p = 2v(e^t - 1), \quad q = v(e^t + 1). \tag{1.11.24}$$

On substituting in the equation for $z$ and integrating, we get

$$z = \frac{5}{2}v^2(e^{2t} - 1) - 3v^2(e^t - 1). \tag{1.11.25}$$

On solving for $t$ and $v$ from (1.11.23), we get

$$e^t = \frac{y - x}{2y - x}, v = x - 2y.$$

On substituting for $t$ and $v$ in terms of $x$ and $y$ in (1.11.25), we get

$$z = \frac{1}{2}y(4x - 3y).$$

In Case (ii), Equations (1.11.22) give $x = s + p, \quad y = q$.
Observe that

$$\frac{d}{dt}(p + q - y) = p + q - y,$$

$$p + q - x = -se^t.$$

Similarly $p + q - y = 0$.
Therefore $p \equiv 0, \quad q = y, \quad x = s, \quad q = y = s - se^t$. Hence

$$s = x, \quad e^t = \frac{x - y}{x},$$

$$\frac{dz}{dt} = y(y - x),$$

$$= s^2(e^{2t} - e^t).$$

On integrating and using the initial conditions, we get

$$z = \frac{s^2}{2}(e^{2t} - 2e^t + 1),$$

$$= \frac{s^2}{2}(e^t - 1)^2.$$

On eliminating $s$ and $t$, we get

$$z = \frac{x^2}{2}\left(\frac{x-y}{x} - 1\right)^2 = \frac{y^2}{2}.\qquad\square$$

**Example 1.11.3**: Find by the method of characteristics, the integral surface of $pq = xy$ that passes through the line $z = x$, $y = 0$.

**Solution**:   Let us find the initial strips. The initial data curve is

$$x_0(s) = s, \quad y_0(s) = 0, \quad z_0(s) = s.$$

In this case, Equations (1.11.18) and (1.11.19) respectively become

$$p_0(s)q_0(s) = 0,$$

and

$$1 = p_0 \cdot 1 + q_0 \cdot 0.$$

Therefore

$$p_0 = 1, q_0 = 0, \qquad \text{(unique initial strip)}.$$

The characteristic equations are

$$\frac{dx}{dt} = q, \ \frac{dy}{dt} = p, \ \frac{dz}{dt} = 2pq, \frac{dp}{dt} = y, \frac{dq}{dt} = x.$$

Hence

$$x = Ae^t + Be^{-t}, \quad q = Ae^t - Be^{-t}.$$

Similarly

$$y = Ce^t + De^{-t}, \quad p = Ce^t - De^{-t}.$$

$$\frac{dz}{dt} = 2pq = 2ACe^{2t} + 2BDe^{-2t} - 2(BC + AD),$$

$$z = ACe^{2t} - BDe^{-2t} - 2(BC + AD)t + E.$$

Using the initial conditions

$$A + B = s, A - B = 0 \Rightarrow A = B = \frac{s}{2},$$

$$C + D = 0, C - D = 1 \Rightarrow C = -D = \frac{1}{2},$$

$$AC - BD + E = s.$$

Therefore $E = \dfrac{s}{2}$.

Finally, $x = s \cosh t, y = \sinh t, z = s \cosh^2 t, p = \cosh t, q = s \sinh t$. Hence the surface passing through the initial data curve is given by $z^2 = x^2(1 + y^2)$.    □

**Example 1.11.4**: Find the characteristic strips of the equation $xp + yq - pq = 0$ and obtain the equation of the integral surface through the curve $C: z = x/2, y = 0$.

**Solution**: The initial data curve is

$$x_0(s) = 2s, \quad y_0(s) = 0, \quad z_0(s) = s.$$

The Equations (1.11.18) and (1.11.19) become respectively

$$2sp_0 - p_0 q_0 = 0 \text{ and } 1 = 2p_0,$$

which implies that

$$p_0 = \frac{1}{2} \text{ and } q_0 = 2s.$$

Since only one initial strip is possible through the given initial data curve, we can find only one solution.

The characteristic equations are

$$\frac{dx}{dt} = x - q, \quad \frac{dy}{dt} = y - p, \quad \frac{dz}{dt} = -pq,$$

$$\frac{dp}{dt} = -p, \quad \frac{dq}{dt} = -q.$$

The characteristic strip issuing from the initial element at 's' on the initial data curve is

$$p = \frac{1}{2}e^{-t}, \quad q = 2se^{-t}, \quad z = \frac{s}{2}(1 + e^{-2t}),$$

$$x = 2s \cosh t, \quad y = -\frac{1}{2}\sinh t.$$

Therefore the integral surface through $C$ is

$$x^2 - 4z^2 + 8xyz = 0. \qquad \square$$

**Exercise 1.11.1**: Find an integral surface of $p^2x + qy - z = 0$ containing the initial line $y = 1$, $x + z = 0$.
**Ans.**: $z = \dfrac{xy}{(y-2)}$ .

**Exercise 1.11.2**: Find the solution of $z = p^2 - q^2$ that passes through the curve $C : x_0 = s, y_0 = 0, z_0 = -\frac{1}{4}s^2$ .
**Hint**: (Two initial strips are possible.)

**Exercise 1.11.3**: Determine the two solutions of the equation $pq = 1$ passing through the straight line $C : x_0 = 2s, y_0 = 2s, z_0 = 5s$.

**Exercise 1.11.4**: Find the characteristic strips of the equation
$$z + px + qy = 1 + pqx^2y^2,$$
passing through the initial data curve $C : x_0 = s, y_0 = 1, z_0 = -s$.

**Exercise 1.11.5**: Find the integral surface of the equation $pq = z$, passing through $C : x_0 = 0, y_0 = s, z_0 = s^2$.
**Ans.**: $z = \dfrac{(x+4y)^2}{16}$ .

**Exercise 1.11.6**: Find the integral surface of the equation $z = p^2 - 3q^2$, passing through $C : x_0 = s, y_0 = 0, z_0 = s^2$. Show that there are two possible initial strips $p_0 = 2s, q_0 = \pm s$.
**Ans.**: For the initial strip $(s, 0, s^2, 2s, s)$, the characteristic strip is $(4s(e^t - 1) + s, -6s(e^t - 1), s^2e^{2t}, 2se^t, se^t)$. The integral surface is given by $z = (2x+y)^2/4$. (Find the integral surface for the other initial strip.)

We will now state and prove that under certain conditions, the solution to the Cauchy problem for a non-linear p.d.e. exists and is unique.

**Theorem 1.11.1**: Consider the p.d.e.

$$f(x, y, z, p, q) = 0, \qquad (1.11.26)$$

where $f$ has continuous second order derivatives with respect to its variables $x, y, z, p,$ and $q$, and at every point either $f_p \neq 0$ or $f_q \neq 0$. Suppose that the initial values $z = z_0(s)$ are specified along the initial curve $\Gamma_0 : x = x_0(s), y = y_0(s), a \leq s \leq b$, where $x_0(s), y_0(s),$ and $z_0(s)$ have continuous second order derivatives. Suppose $p_0(s)$ and $q_0(s)$ have been determined such that

$$f(x_0(s), y_0(s), z_0(s), p_0(s), q_0(s)) = 0,$$

and

$$\frac{dz_0}{ds} = p_0 \frac{dx_0}{ds} + q_0 \frac{dy_0}{ds} \, ,$$

where $p_0$ and $q_0$ are continuously differentiable functions of $s$. If, in addition, the five functions $x_0, y_0, z_0, p_0$, and $q_0$ satisfy

$$f_q \frac{dx_0}{ds} - f_p \frac{dy_0}{ds} \neq 0,$$

then in some neighborhood of each point of the initial curve there exists one and only one solution $z = z(x, y)$ of (1.11.26) such that

$$z(x_0(s), y_0(s)) = z_0(s),$$
$$z_x(x_0(s), y_0(s)) = p_0(s),$$
$$z_y(x_0(s), y_0(s)) = q_0(s).$$

That is, the integral surface $z = z(x, y)$ contains the initial strip.
**Proof**: Let us consider the system of characteristic equations

$$\frac{dx}{dt} = f_p \, , \quad \frac{dy}{dt} = f_q \, , \quad \frac{dz}{dt} = pf_p + qf_q,$$

$$\frac{dp}{dt} = -(f_x + pf_z) \, , \quad \frac{dq}{dt} = -(f_y + qf_z),$$

with the initial conditions $x = x_0(s)$, $y = y_0(s)$, $z = z_0(s)$, $p = p_0(s)$, $q = q_0(s)$ at $t = 0$. We can solve this system of ordinary differential equations to obtain (from the existence and uniqueness theorem for the initial value problems for a system of ordinary differential equations)

$$x = X(s,t), \quad y = Y(s,t), \quad z = Z(s,t), \quad p = P(s,t), \quad q = Q(s,t),$$

where $X, Y, Z, P$, and $Q$ have continuous derivatives with respect to $s$ and $t$ and such that they satisfy the initial conditions

$$X(s,0) = x_0(s), \quad Y(s,0) = y_0(s), \quad Z(s,0) = z_0(s),$$

$$P(s,0) = p_0(s), \quad Q(s,0) = q_0(s).$$

**Proposition 1.11.1**: The characteristic curves $x = X(s,t)$, $y = Y(s,t), z = Z(s,t)$ indeed form an integral surface. That is, if $s$ and $t$ are solved in terms of $x$ and $y$

from the first two equations, then the last equation gives the solution expressed in the form $z = Z(s(x, y), t(x, y)) = z(x, y)$.

Let $(x_0, y_0)$ be a point on the initial curve. Since the Jacobian

$$\left.\frac{\partial(X, Y)}{\partial(s, t)}\right|_{t=0} = f_q \frac{dx_0}{ds} - f_p \frac{dy_0}{ds} \neq 0,$$

at every point on the initial curve there exists a neighborhood $N(x_0, y_0)$ such that the Jacobian $\dfrac{\partial(X, Y)}{\partial(s, t)} \neq 0$. In this neighborhood $N(x_0, y_0)$, we have the following well defined functions.

$$\begin{aligned}
s &= s(x, y), x \equiv X(s(x, y), t(x, y)), \\
t &= t(x, y), y \equiv Y(s(x, y), t(x, y)), \\
z &= Z(s(x, y), t(x, y)) = z(x, y), \\
p &= P(s(x, y), t(x, y)) = p(x, y), \\
q &= Q(s(x, y), t(x, y)) = q(x, y).
\end{aligned} \qquad (1.11.27)$$

We will now show that $z(x, y)$ satisfies

$$f(x, y, z(x, y), z_x(x, y), z_y(x, y)) = 0.$$

It suffices to show that $p = z_x$ and $q = z_y$ since $f(x, y, z, p, q) = 0$ as the characteristic strips are essentially solutions of $(1.11.11)$. Let us consider

$$U(s, t) = Z_s - PX_s - QY_s.$$

At $t = 0, U(s, 0) = \dfrac{dz_0}{ds} - p_0 \dfrac{dx_0}{ds} - q_0 \dfrac{dy_0}{ds} = 0$, by the strip condition for the initial elements.

We shall show that $U = 0$ for all $t$.

Consider

$$\begin{aligned}
\frac{\partial U}{\partial t} &= Z_{st} - P_t X_s - Q_t Y_s - PX_{st} - QY_{st}, \\
&= \frac{\partial}{\partial s}(Z_t - PX_t - QY_t) + P_s X_t + Q_s Y_t - Q_t Y_s - P_t X_s, \\
&= 0 + P_s f_p + Q_s f_q + (f_x + f_z P)X_s + (f_y + f_z Q)Y_s,
\end{aligned}$$

by making use of the strip condition for characteristics. This can be rewritten as

$$\begin{aligned}
\frac{\partial U}{\partial t} &= f_x X_s + f_y Y_s + f_z Z_s + f_p P_s + f_q Q_s - f_z(Z_s - PX_s - QY_s),\\
&= f_s - f_z U,\\
&= -f_z U, \qquad \text{(since } f \equiv 0 \ \forall \ s \text{ and } t\text{)}.
\end{aligned}$$

Therefore, for a fixed $s$ (i.e., along the characteristic curve), $U$ satisfies the following ordinary differential equation

$$\frac{dU}{dt} = -f_z U,$$

which has a solution

$$U = U(0)\exp\left(-\int_0^t f_z dt\right).$$

Combined with the condition that $U = 0$ for $t = 0$, the previous expression implies that $U \equiv 0$ for all $t$. Therefore

$$Z_s = PX_s + QY_s. \tag{1.11.28}$$

Consider the four equations

$$\text{(i)} \quad Z_s = PX_s + QY_s, \qquad \text{(ii)} \quad Z_t = PX_t + QY_t,$$

$$\text{(iii)} \ Z_s = z_x X_s + z_y Y_s, \qquad \text{(iv)} \quad Z_t = z_x X_t + z_y Y_t.$$

(i) has been proved (refer to Equation (1.11.28)). (ii) is the characteristic equation. (iii) and (iv) are obtained by differentiating the identities of Equation (1.11.27).

Observe that $(P, Q)$ satisfy (i) and (ii) and $(z_x, z_y)$ satisfy (iii) and (iv). As a system of simultaneous equations (i) and (ii) are the same as (iii) and (iv). Further, the determinant of the coefficients of these linear systems is $\dfrac{\partial(X, Y)}{\partial(s, t)}$. This determinant is non-zero in the neighborhood $N(x_0, y_0)$. Hence we have

$$P(s, t) = z_x(x(s, t), y(s, t)), \ \ Q(s, t) = z_y(x(s, t), y(s, t)),$$

or

$$p(x, y) = z_x(x, y), \ \ q(x, y) = z_y(x, y),$$

which was to be shown. Hence the result.

Also, the solution $z = z(x, y)$ contains the initial strip. For,

$$z(x_0, y_0) = z(x(s, 0), y(s, 0)) = Z(s, 0) = z_0(s),$$
$$z_x(x_0, y_0) = p(x_0, y_0) = p(x(s, 0), y(s, 0)) = P(s, 0) = p_0(s),$$
$$z_y(x_0, y_0) = q(x_0, y_0) = q(x(s, 0), y(s, 0)) = Q(s, 0) = q_0(s).$$

**Proposition 1.11.2**:   $z = z(x, y)$ as determined in Proposition (1.11.1) is unique. Suppose instead that there are two solutions, say, $z = z_1(x, y)$ and $z = z_2(x, y)$ containing the initial strip. Consider any point on the initial data curve. These two surfaces contain the initial element at that point, and hence the characteristic strip issuing from this initial element through that point (refer to Lemma 1.11.1). Therefore the corresponding characteristic curve has to lie on both the surfaces. As this happens for every point on the initial data curve, both these surfaces are generated by the same set of characteristic curves and must hence coincide. Thus the solution is unique.                                                    □

# Chapter 2

# Second Order Partial Differential Equations

## 2.1   Genesis of Second Order P.D.E.

**Definition 2.1.1**: A partial differential equation is said to be a second order semi-linear p.d.e. if it can be put in the form

$$R(x,y)u_{xx} + S(x,y)u_{xy} + T(x,y)u_{yy} + g(x,y,u,u_x,u_y) = 0, \qquad (2.1.1)$$

where $R^2 + S^2 + T^2 \neq 0$ and $R, S$, and $T$ are continuous functions of $x$ and $y$.

**Definition 2.1.2**:   A function $u = u(x,y)$ is said to be a regular solution of (2.1.1) in a domain $D \subseteq \mathbb{R} \times \mathbb{R}$ if $u \in C^2(D)$, and the function and its derivatives satisfy Equation (2.1.1) identically in $x$ and $y$ for $(x,y) \in D$.

**Example 2.1.1**: If $f \in C^2(\mathbb{R})$ then $u = f(x+t)$ is a solution of the second order p.d.e.

$$u_{xx} - u_{tt} = 0. \qquad \square$$

We now discuss some examples of second order partial differential equations that arise in physics and mathematics.

**Transverse vibrations of a string**:   Let $y = y(x,t)$ be the transverse displacement from the mean position ($x$-axis) of a string at time $t$ at the point $x$. Consider a small portion of the string $\Delta s$ between the two points $P$ and $Q$. The forces acting on this part of the string are tensions at $P$ and $Q$. We neglect the weight of the string. The equations of motion are (refer to Fig. 2.1.1)

(i)  in the $x$-direction (assuming no displacement in the $x$ direction)

$$T_2 \cos \psi_2 = T_1 \cos \psi_1 = T \quad \text{(say)},$$

(ii) in the $y$-direction

$$
\begin{aligned}
(\rho \Delta s) y_{tt} &= T_2 \sin \psi_2 - T_1 \sin \psi_1 , \\
&= T(\tan \psi_2 - \tan \psi_1) ,
\end{aligned}
$$

where $\rho$ is the linear density of the string.

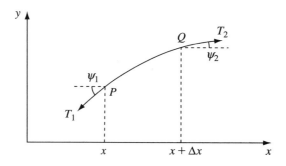

Figure 2.1.1

Observe

$$\tan \psi_1 = (y_x)|_P, \quad \tan \psi_2 = (y_x)|_Q \simeq (y_x)|_P + (y_{xx})|_P \Delta x,$$

where $|_P$ and $|_Q$ indicate the evaluation at the points $P$ and $Q$ respectively. Hence

$$(\rho \Delta s) y_{tt} = T[(y_x)|_P + (y_{xx})|_P \Delta x - (y_x)|_P].$$

Taking the limit as $\Delta s \to 0$ and $\Delta x \to 0$ after dividing both sides of the previous equation by $\Delta x$, we obtain

$$\rho y_{tt} = \frac{T y_{xx}}{\sqrt{1 + y_x^2}} .$$

If $| y_x | \ll 1$ (i.e., if the slope is small everywhere), we get

$$y_{xx} = \frac{1}{c^2} y_{tt} , \quad c^2 = \frac{T}{\rho} .$$

The previous equation is called the linear one-dimensional wave equation.

**Heat conduction equation**: Let us consider a homogeneous, isotropic solid. (Homogeneous means that the material properties are translational invariant and isotropic means that the material properties are the same in all directions.) Let $V$ be any arbitrary volume inside the solid bounded by a surface $S$. Let $\delta V$ be a volume element. The heat energy stored in $\delta V$ is equal to $c\rho u \delta V$, where $c$ is the specific heat of the solid, $\rho$ is its density, and $u$ is the temperature that is a function of position and time. Therefore

$$\text{the total heat energy in } V = \iiint_V c\rho u \, dV.$$

Let $\delta S$ be a surface element. The heat flow across $\delta S = k\nabla u \cdot \vec{n}\delta S$ where $\vec{n}$ is the outward drawn normal to the surface $S$ and $k$ is the thermal conductivity of the solid. Therefore, the total flux across $S$, using the divergence theorem, is given as follows

$$\iint_S k\,\nabla u \cdot \vec{n}\,dS = \iiint_V \nabla \cdot (k\,\nabla u)\,dV.$$

The rate of change of heat energy in $V$ = the flux of heat energy across $S$, i.e.,

$$\frac{d}{dt}\iiint_V c\rho u\,dV = \iiint_V \nabla \cdot (k\nabla u)\,dV,$$

$$\iiint_V \left(\frac{\partial(c\rho u)}{\partial t} - \nabla \cdot (k\nabla u)\right)dV = 0.$$

As $V$ is arbitrary and the integrand is continuous, we have

$$c\rho\frac{\partial u}{\partial t} - \nabla \cdot (k\,\nabla u) = 0.$$

If the conductivity $k$ is constant throughout the body, then

$$\frac{\partial u}{\partial t} = \kappa\nabla^2 u,$$

which is referred to as the heat conduction equation, where $\kappa = k/c\rho$. In the one-dimensional case, this equation becomes

$$\frac{\partial u}{\partial t} = \kappa\frac{\partial^2 u}{\partial x^2}\,.$$

**Analytic function of** $z$: If $f(z) = u(x,y) + iv(x,y)$ is analytic in $z(= x+iy)$, then it is well known that $u$ and $v$ satisfy the Cauchy-Riemann equations

$$u_x = v_y\,,\ u_y = -v_x\,.$$

Therefore $\qquad\qquad\qquad u_{xx} + u_{yy} = 0.$

Similarly $\qquad\qquad\qquad v_{xx} + v_{yy} = 0.$

These equations are called two-dimensional Laplace's equations.

Second order partial differential equations widely occur in many other areas of physics, chemistry, mathematics, and engineering.

## 2.2    Classification of Second Order P.D.E.

Consider the following second order semi-linear p.d.e.

$$Lu + g(x, y, u, u_x, u_y) = 0, \qquad\qquad (2.2.1)$$

where

$$L = R(x, y)\frac{\partial^2}{\partial x^2} + S(x, y)\frac{\partial^2}{\partial x \partial y} + T(x, y)\frac{\partial^2}{\partial y^2},$$

$R^2 + S^2 + T^2 \neq 0$ and $R, S,$ and $T$ are continuous functions of $x$ and $y$. We further assume that these functions possess continuous partial derivatives with respect to $x$ and $y$.

We change the independent variables $(x, y)$ to $(\xi, \eta)$

$$\xi = \xi(x, y), \quad \eta = \eta(x, y).$$

**Note**: We assume that $(\xi_x \eta_y - \xi_y \eta_x) \neq 0$ so that the transformation is invertible at least locally.

It is easy to see that

$$
\begin{aligned}
u_x &= u_\xi \xi_x + u_\eta \eta_x \ , \\
u_y &= u_\xi \xi_y + u_\eta \eta_y \ , \\
u_{xy} &= u_{\xi\xi} \xi_x \xi_y + u_{\xi\eta} \xi_x \eta_y + u_\xi \xi_{xy} \\
&\quad + u_{\eta\xi} \eta_x \xi_y + u_{\eta\eta} \eta_x \eta_y + u_\eta \eta_{xy} \ , \\
u_{xx} &= u_{\xi\xi} \xi_x^2 + u_{\xi\eta} \xi_x \eta_x + u_\xi \xi_{xx} \\
&\quad + u_{\eta\xi} \eta_x \xi_x + u_{\eta\eta} \eta_x^2 + u_\eta \eta_{xx} \ , \\
u_{yy} &= u_{\xi\xi} \xi_y^2 + u_{\xi\eta} \xi_y \eta_y + u_\xi \xi_{yy} \\
&\quad + u_{\eta\xi} \eta_y \xi_y + u_{\eta\eta} \eta_y^2 + u_\eta \eta_{yy} \ .
\end{aligned}
$$

Therefore

$$Ru_{xx} + Su_{xy} + Tu_{yy} = u_{\xi\xi}(R\xi_x^2 + S\xi_x\xi_y + T\xi_y^2)$$
$$+ u_{\xi\eta}[2R\xi_x\eta_x + S(\xi_x\eta_y + \eta_x\xi_y) + 2T\xi_y\eta_y]$$
$$+ u_{\eta\eta}(R\eta_x^2 + S\eta_x\eta_y + T\eta_y^2)$$
$$+\text{a function of } (\xi, \eta, u_\xi, u_\eta, u).$$

Therefore Equation (2.2.1) becomes

$$A(\xi_x, \xi_y)u_{\xi\xi} + 2B(\xi_x, \xi_y; \eta_x, \eta_y)u_{\xi\eta} + A(\eta_x, \eta_y)u_{\eta\eta} = G(\xi, \eta, u, u_\xi, u_\eta), \qquad (2.2.2)$$

where

$$A(u, v) = Ru^2 + Suv + Tv^2, \qquad (2.2.3a)$$

$$B(u_1, v_1; u_2, v_2) = Ru_1u_2 + \frac{1}{2}S(u_1v_2 + u_2v_1) + Tv_1v_2. \qquad (2.2.3b)$$

**Exercise 2.2.1**: Show that

$$A(\xi_x, \xi_y)A(\eta_x, \eta_y) - B^2(\xi_x, \xi_y; \eta_x, \eta_y) = (4RT - S^2)(\xi_x\eta_y - \xi_y\eta_x)^2/4. \qquad (2.2.4)$$

The problem now is to choose $\xi$ and $\eta$ so that Equation (2.2.2) takes a simple form.
**Case (i)**: $S^2 - 4RT > 0$. $\qquad\qquad\qquad\qquad\qquad\qquad\qquad\qquad\qquad (2.2.5)$
Then we shall show that $\xi$ and $\eta$ can be so chosen that the coefficients of $u_{\xi\xi}$ and $u_{\eta\eta}$ in (2.2.2) vanish.
Consider $R\alpha^2 + S\alpha + T = 0$.
This equation has two real distinct roots $\lambda_1(x, y)$ and $\lambda_2(x, y)$ due to the condition (2.2.5).
We choose $\xi$ and $\eta$ such that

$$\frac{\partial \xi}{\partial x} = \lambda_1 \frac{\partial \xi}{\partial y}, \quad \frac{\partial \eta}{\partial x} = \lambda_2 \frac{\partial \eta}{\partial y}.$$

These are first order partial differential equations for $\xi$ and $\eta$. If $f_1(x, y) = c_1$ and $f_2(x, y) = c_2$ are the solutions of the ordinary differential equations

$$\frac{dy}{dx} + \lambda_1(x, y) = 0, \quad \frac{dy}{dx} + \lambda_2(x, y) = 0,$$

respectively, then

$$\xi = f_1(x, y), \quad \eta = f_2(x, y),$$

will be the suitable choice. This choice of $\xi$ and $\eta$ makes

$$A(\xi_x, \xi_y) = A(\eta_x, \eta_y) = 0.$$

From Equation (2.2.4), in this case, we have $B^2 > 0$. Hence Equation (2.2.2) reduces to

$$\frac{\partial^2 u}{\partial \xi \partial \eta} = \phi(\xi, \eta, u, u_\xi, u_\eta). \qquad (2.2.6)$$

The curves $\xi(x, y) = $ constant and $\eta(x, y) = $ constant, are called the characteristic curves of Equation (2.2.1).

**Case (ii)**: $S^2 - 4RT = 0$. $\qquad (2.2.7)$

In this case, the roots of the equation $R\alpha^2 + S\alpha + T = 0$ coincide (say, $\lambda(x, y)$). Define $\xi = f(x, y)$, where $f(x, y) = c_1$ is the solution of

$$\frac{dy}{dx} + \lambda(x, y) = 0.$$

Take $\eta$ as any arbitrary function of $x$ and $y$ independent of $\xi$ and from (2.2.4) observe that $B = 0$ since $A(\xi_x, \xi_y) = 0$.

Since $\eta$ is independent of $\xi$, $A(\eta_x, \eta_y) \neq 0$. Therefore Equation (2.2.2) reduces to

$$\frac{\partial^2 u}{\partial \eta^2} = \phi(\xi, \eta, u, u_\xi, u_\eta). \qquad (2.2.8)$$

**Case (iii)**: $S^2 - 4RT < 0$. $\qquad (2.2.9)$

This is formally the same as Case (i). However here the roots are complex. Proceeding as in Case (i), we find that Equation (2.2.2) reduces to the form (2.2.6) but that the variables $\xi$ and $\eta$ are not real and are in fact complex conjugates. Therefore there are no real characteristic curves in this case.

We make the further transformation

$$\alpha = \frac{1}{2}(\xi + \eta), \quad \beta = \frac{i}{2}(\eta - \xi).$$

Equation (2.2.2) then becomes

$$\frac{\partial^2 u}{\partial \alpha^2} + \frac{\partial^2 u}{\partial \beta^2} = \phi(\alpha, \beta, u, u_\alpha, u_\beta). \qquad (2.2.10)$$

### Classification of Second Order Partial Differential Equations

(i)   $S^2 - 4RT > 0$ Hyperbolic type Eg: Wave equation.
(ii)  $S^2 - 4RT = 0$ Parabolic type   Eg: Heat conduction equation.
(iii) $S^2 - 4RT < 0$ Elliptic type    Eg: Laplace's equation.

**Canonical forms**: Equations (2.2.6), (2.2.8), and (2.2.10) are said to be the canonical forms of the hyperbolic, parabolic, and elliptic type of equations respectively.

**Example 2.2.1**: Reduce the equation $u_{xx} - x^2 u_{yy} = 0$ to a canonical form.

**Solution**: In this case, $R = 1, S = 0, T = -x^2$.

Then $S^2 - 4RT = 4x^2 > 0$ (hyperbolic type).

$R\alpha^2 + S\alpha + T = 0$ becomes $\alpha^2 \quad x^2 = 0$. This gives

$$\alpha = \pm x \quad \Rightarrow \lambda_1 = x, \ \lambda_2 = -x.$$

Observe that

$$y + \frac{x^2}{2} = c_1,$$

and

$$y - \frac{x^2}{2} = c_2,$$

are the solutions of

$$\frac{dy}{dx} \pm x = 0.$$

Therefore $\xi = y + \dfrac{x^2}{2}$ and $\eta = y - \dfrac{x^2}{2}$. Changing the independent variables from $x$ and $y$ to $\xi$ and $\eta$, we obtain

$$u_x = u_\xi x - u_\eta x ,$$
$$u_y = u_\xi + u_\eta ,$$
$$u_{xx} \doteq x^2 u_{\xi\xi} - 2x^2 u_{\xi\eta} + x^2 u_{\eta\eta} + u_\xi - u_\eta ,$$
$$u_{yy} = u_{\xi\xi} + 2u_{\xi\eta} + u_{\eta\eta} .$$

Therefore the given equation becomes

$$u_{\xi\eta} = \frac{1}{4x^2}(u_\xi - u_\eta) = \frac{1}{4(\xi - \eta)}(u_\xi - u_\eta). \qquad \square$$

**Example 2.2.2**: Reduce the equation

$$y^2 u_{xx} - 2xy u_{xy} + x^2 u_{yy} = \frac{y^2}{x} u_x + \frac{x^2}{y} u_y ,$$

to a canonical form and solve it.

**Solution**: In this case, $S^2 - 4RT = 0$ (parabolic type).
The equation $R\alpha^2 + S\alpha + T = 0$ becomes $y^2 \alpha^2 - 2xy\alpha + x^2 = 0$.

$$(y\alpha - x)^2 = 0 \Rightarrow \lambda = \frac{x}{y} .$$

A solution of

$$\frac{dy}{dx} + \frac{x}{y} = 0,$$

is $x^2 + y^2 = c_1$. Therefore

$$\xi = x^2 + y^2.$$

Let us choose

$$\eta = x^2 - y^2.$$

Observe that $\eta$ is independent of $\xi$, for, $\dfrac{\partial(\xi, \eta)}{\partial(x, y)} \neq 0$. The given equation transforms to

$$u_{\eta\eta} = 0.$$

Hence

$$u_\eta = f(\xi),$$

which implies that

$$u = f(\xi)\eta + g(\xi),$$

i.e.,    $$u(x, y) = f(x^2 + y^2)(x^2 - y^2) + g(x^2 + y^2). \qquad \square$$

**Example 2.2.3**: Reduce the equation $u_{xx} + x^2 u_{yy} = 0$ to a canonical form.
**Solution**: In this case, $S^2 - 4RT = -4x^2 < 0$ (elliptic type).
The equation $R\alpha^2 + S\alpha + T = 0$ becomes $\alpha^2 + x^2 = 0$. This implies that

$$\lambda_1 = ix, \lambda_2 = -ix.$$

Here

$$\xi = iy + \frac{x^2}{2} , \ \eta = -iy + \frac{x^2}{2} .$$

Therefore

$$\alpha = \frac{x^2}{2} \ , \beta = y \ .$$

The given equation then transforms to the canonical form

$$u_{\alpha\alpha} + u_{\beta\beta} = -\frac{1}{2\alpha}u_\alpha. \qquad \Box$$

**Example 2.2.4**: Reduce the equation

$$(n-1)^2 u_{xx} - y^{2n} u_{yy} = ny^{2n-1} u_y,$$

where $n$ is an integer to a canonical form.

**Solution**: If $n = 1$, the equation reduces to $u_{yy} = -\frac{1}{y}u_y$, which is in the canonical form (parabolic). Then

$$yu_{yy} + u_y = 0 \Rightarrow \frac{\partial(yu_y)}{\partial y} = 0.$$

Thus $yu_y = f(x)$, where $f(x)$ is an arbitrary function. Therefore we have $u_y = \frac{1}{y}f(x)$, and thus $u = f(x)\log y + g(x)$, where $g(x)$ is an arbitrary function.

If $n > 1$, then it is hyperbolic in type. Here $\xi = x + y^{1-n}$ and $\eta = x - y^{1-n}$. Then the given equation transforms to the canonical form

$$u_{\xi\eta} = 0.$$

The solution is $u(x, y) = f(x + y^{1-n}) + g(x - y^{1-n})$.

**Note**: The arbitrary functions $f$ and $g$ that occur in the solution in Examples 2.2.2 and 2.2.4 have to be $C^2$ functions so that the solution is a regular solution. $\qquad \Box$

**Exercise 2.2.2**: Reduce the following into canonical forms and solve whenever possible.

(i) $u_{xx} + 2u_{xy} + 17u_{yy} = 0$.

(ii) $x^{14}u_{xx} - 36u_{yy} + 7x^{13}u_x = 0$.

(iii) $u_{xx} - 4x^2 u_{yy} = \frac{1}{x}u_x$.

(iv) $u_{xx} - (\text{sech}^4 x)\, u_{yy} = 0$.

(v) $u_{xx} + xu_{yy} = 0$ in the region $x < 0$.

(vi) $xu_{xx} + 2\sqrt{xy}u_{xy} + yu_{yy} - u_x = 0$.

(vii) $e^{2x}u_{xx} + 2e^{x+y}u_{xy} + e^{2y}u_{yy} = 0$.

(viii) $(\sin^2 x)u_{xx} + 2(\cos x)u_{xy} - u_{yy} = 0$.

(ix) $x^2(y-1)u_{xx} - x(y^2-1)u_{xy} + y(y-1)u_{yy} + xyu_x - u_y = 0$.

## 2.3    One Dimensional Wave Equation

In Sections 2.3.1 to 2.3.3 and Section 2.3.5, we will use $y$ as a dependent variable with $x$ and $t$ as independent variables.

### 2.3.1    Vibrations of an Infinite String

Consider the following one-dimensional wave equation

$$\frac{\partial^2 y}{\partial x^2} = \frac{1}{c^2}\frac{\partial^2 y}{\partial t^2}, \qquad -\infty < x < \infty, \ \ t > 0. \tag{2.3.1}$$

If we introduce the characteristic variables

$$\xi = x - ct, \quad \eta = x + ct,$$

Equation (2.3.1) transforms to

$$\frac{\partial^2 y}{\partial \xi \partial \eta} = 0.$$

Therefore

$$y(x,t) = F(\xi) + G(\eta),$$
$$= F(x - ct) + G(x + ct).$$

We need the arbitrary functions $F$ and $G$ to be $C^2$ functions so that $y(x,t)$ is a regular solution of (2.3.1). The previous solution is indeed the general solution of (2.3.1).

Let the initial conditions be given as follows

$$y(x,0) = f(x), \quad y_t(x,0) = g(x), \quad -\infty < x < \infty,$$

where $y = f(x)$ is the initial position of the string and $g(x)$ is the initial velocity at the point $x$. Hence

$$F(x) + G(x) = f(x),$$
$$-cF'(x) + cG'(x) = g(x).$$

Solving for $F(x)$ and $G(x)$ we get,

$$F(x) = \frac{1}{2c}[cf(x) - \int_{x_0}^{x} g(x)dx],$$

$$G(x) = \frac{1}{2c}[cf(x) + \int_{x_0}^{x} g(x)dx].$$

Therefore

$$y(x, t) = \frac{1}{2}[f(x - ct) + f(x + ct)] + \frac{1}{2c}\int_{x-ct}^{x+ct} g(s)ds, \qquad (2.3.2)$$

where we require $f \in C^2$ and $g \in C^1$ so that $y(x, t)$ is a $C^2$ function. This is called **d'Alembert's solution**, which describes the vibrations of an infinite string.

**Note**:

1. $f(x - ct)$ is a wave form travelling with speed $c$ along $ox$. (What is a wave?)

2. $x - ct = $ constant and $x + ct = $ constant, are the characteristic curves of Equation (2.3.1).

Properties of characteristics

(a) **Characteristics are carriers of discontinuities**:
Let the initial shape be triangular as shown in Fig. 2.3.1 and given by

$$f(x) = \begin{cases} 0, & -\infty < x \leq -a, \\ x + a, & -a \leq x \leq 0, \\ a - x, & 0 \leq x \leq a, \\ 0. & a \leq x < \infty. \end{cases}$$

Suppose the initial velocity is given by $g(x) = 0$. Then Equation (2.3.2) becomes

$$y(x, t) = \frac{1}{2}[f(x - ct) + f(x + ct)]. \qquad (2.3.3)$$

Observe that $f$ is a continuous function in $(-\infty, \infty)$ but is not differentiable at $x = 0, \pm a$. Hence the initial shape of the string (Fig. 2.3.1) has kinks at these points. We shall see how these kinks propagate.

In this case, (2.3.3) is not a regular solution as it is not differentiable twice. (Such solutions are called 'weak solutions'.)

Observe that the initial shape divides into two equal halves and each travels with a speed $c$ in opposite directions. In general, there are six jumps in the slope.

The loci of these six points where there is a jump in the slope are as follows.

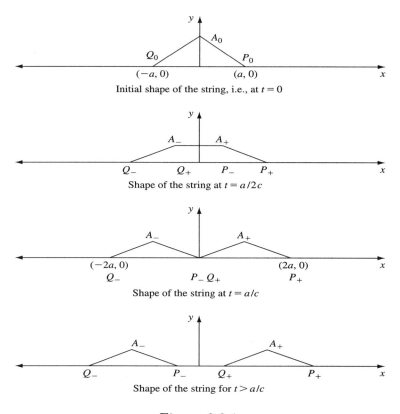

Figure 2.3.1

$$x - ct = a \quad \text{(locus of } P^+),$$
$$x - ct = 0 \quad \text{(locus of } A^+),$$
$$x - ct = -a \quad \text{(locus of } Q^+),$$
$$x + ct = a \quad \text{(locus of } P^-),$$
$$x + ct = 0 \quad \text{(locus of } A^-),$$
$$x + ct = -a \quad \text{(locus of } Q^-).$$

**(b)  Characteristics determine the solution uniquely only if the data is prescribed on both of them**:

Consider the general solution of wave equation

$$y = F(\xi) + G(\eta) = F(x - ct) + G(x + ct).$$

Then
$$y_t = -cF'(x - ct) + cG'(x + ct).$$

Let the data be given on $\xi = 0$, i.e., $y(0, \eta) = g(\eta)$.
Therefore
$$g(\eta) = F(0) + G(\eta).$$

Therefore
$$y = F(\xi) - F(0) + g(\eta),$$
$$y_t(0, \eta) = -cF'(0) + cg'(\eta),$$

and hence $y_t$ cannot be prescribed arbitrarily on $\xi = 0$. Moreover, we do not have any condition to determine $F(\xi)$, which is left arbitrary and thus the solution is not unique.

So the characteristics are curves such that if the data is prescribed only on one of them, the solution is not unique.

However, if we prescribe the data on both $\xi = 0$ and $\eta = 0$, the solution is unique, i.e.,
$$y(0, \eta) = g(\eta), \quad y(\xi, 0) = f(\xi).$$

Then
$$g(\eta) = F(0) + G(\eta),$$
$$f(\xi) = F(\xi) + G(0).$$

Therefore
$$y(\xi, \eta) = f(\xi) - G(0) + g(\eta) - F(0),$$
$$= f(\xi) + g(\eta) - f(0), \text{ since} f(0) = g(0).$$

Thus the solution is uniquely determined if the data is given on both the characteristics.

(c)  **Characteristics determine the 'domain of dependence' and the 'range of influence':**

Let $P(x_1, t_1)$ be any point with $t_1 > 0$. From (2.3.2), we have
$$y(x_1, t_1) = \frac{1}{2}[f(x_1 - ct_1) + f(x_1 + ct_1)] + \frac{1}{2c} \int_{x_1 - ct_1}^{x_1 + ct_1} g(z)dz,$$
$$y(x_1, t_1) = \frac{1}{2}[f(A) + f(B)] + \frac{1}{2c} \int_A^B g(z)dz.$$

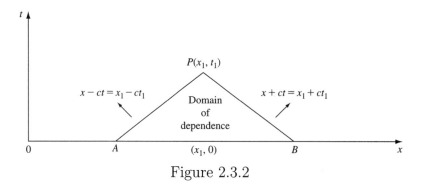

Figure 2.3.2

Hence $y(P)$ depends only on the data given on the line segment $AB$ (Fig. 2.3.2). For this reason, the segment $AB$ is called the '**domain of dependence**' for $P$. The segment $AB$ is determined by the two characteristics passing through $P$. In fact, any point lying inside the triangular region bounded by the two characteristics and the segment $AB$ will have its domain of dependence within the segment $AB$.

The data at $(x_1, 0)$ on $ox$ will influence the values of $y(x, t)$ at any point $P$ lying in the region bounded by the two characteristics passing through $A(x_1, 0)$ (Fig. 2.3.3). This region is called the '**range of influence**' of the point $(x_1, 0)$.

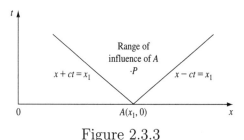

Figure 2.3.3

Observe that any point $P$ that lies in this sector has the point $A$ within its domain of dependence.

## 2.3.2    Vibrations of a Semi-infinite String

Consider the motion of a semi-infinite string (i.e., one end $x = 0$ is kept fixed). The equation governing the motion of the string is

$$\frac{\partial^2 y}{\partial x^2} = \frac{1}{c^2} \frac{\partial^2 y}{\partial t^2}, \quad 0 < x < \infty, \ t > 0.$$

The initial conditions are

$$y(x,0) = u(x), \quad y_t(x,0) = v(x), \quad x \geq 0.$$

The boundary condition is

$$y(0,t) = 0, \quad t \geq 0 \text{ (i.e., the end } x = 0 \text{ is fixed).}$$

Observe that the previous condition implies

$$y_t(0,t) = 0.$$

However, d'Alembert's solution (2.3.2) cannot be used for the initial value problem stated previously, since $u(x - ct)$ has no meaning for the values $t > (x/c)$. Instead we consider the modified problem for an infinite string as follows

$$y(x,0) = f(x), \quad y_t(x,0) = g(x), \quad -\infty < x < \infty,$$

where

$$f(x) = \begin{cases} u(x), & \text{if } x \geq 0, \\ -u(-x), & \text{if } x \leq 0, \end{cases}$$

$$g(x) = \begin{cases} v(x), & \text{if } x \geq 0, \\ -v(-x), & \text{if } x \leq 0. \end{cases}$$

Observe that $f(x)$ and $g(x)$ are defined as odd functions of $x$ so as to satisfy the boundary conditions. Then d'Alembert's solution (2.3.2) becomes

$$y(x,t) = \frac{1}{2}[f(x - ct) + f(x + ct)] + \frac{1}{2c}\int_{x-ct}^{x+ct} g(s)ds.$$

When $x = 0$,

$$y(0,t) = \frac{1}{2}[f(-ct) + f(ct)] + \frac{1}{2c}\int_{-ct}^{ct} g(s)ds,$$

$$y_t(0,t) = \frac{1}{2}[-cf'(-ct) + cf'(ct)] + \frac{1}{2}[g(ct) + g(-ct)].$$

Hence, by making use of the fact that $f$ and $g$ are odd functions, we get

$$y(0,t) = \frac{1}{2}[u(ct) - u(ct)] = 0.$$

$$y_t(0,t) = \frac{c}{2}[u'(ct) - u'(ct)] + \frac{1}{2}[g(ct) + g(-ct)] = 0.$$

At $t = 0$ and $x > 0$,

$$y(x, 0) = \frac{1}{2}[f(x) + f(x)] = \frac{1}{2}[u(x) + u(x)] = u(x).$$

$$y_t(x, 0) = \frac{1}{2}[cf'(x) - cf'(x)] + \frac{1}{2}[g(x) + g(x)] = v(x).$$

Thus the initial and the boundary conditions are satisfied. If the string is released from rest, i.e., if $v \equiv 0$ the solution becomes

$$y(x, t) = \begin{cases} \dfrac{1}{2}[u(x - ct) + u(x + ct)], & x \geq ct, \\[3mm] \dfrac{1}{2}[u(x + ct) - u(ct - x)], & x \leq ct. \end{cases}$$

## 2.3.3   Vibrations of a String of Finite Length

We now consider the vibrations of a string of finite length $l$. The initial conditions are

$$y(x, 0) = u(x), \quad y_t(x, 0) = v(x), \quad 0 \leq x \leq l.$$

The boundary conditions are

$$\left. \begin{array}{l} y(0, t) = 0, \\ y(l, t) = 0, \end{array} \right\} t > 0.$$

This case can also be deduced from d'Alembert's solution by converting this problem into a problem of vibrations of an infinite string. For this purpose, we extend the given data as follows. Define

$$f(x) = \begin{cases} u(x), & \text{if } 0 \leq x \leq l, \\ -u(-x), & \text{if } -l \leq x \leq 0, \end{cases}$$

$$f(x + 2kl) = f(x), \quad -l \leq x \leq l, \quad k = \pm 1, \pm 2, \cdots.$$

Observe that $f(x)$ is an odd function and is periodic with period $2l$. Similarly $g(x)$ can be defined as a periodic odd function with period $2l$ and coinciding with $v(x)$ in $0 \leq x \leq l$.

Let us assume that $f(x)$ can be expanded as a Fourier series. Then

$$f(x) = \sum_{m=1}^{\infty} u_m \sin\left(\frac{m\pi x}{l}\right),$$

where

$$u_m = \frac{2}{l} \int_0^l u(s) \sin\left(\frac{m\pi s}{l}\right) ds.$$

Similarly

$$g(x) = \sum_{m=1}^{\infty} v_m \sin\left(\frac{m\pi x}{l}\right),$$

where

$$v_m = \frac{2}{l} \int_0^l v(s) \sin\left(\frac{m\pi s}{l}\right) ds.$$

Therefore

$$\frac{1}{2}[f(x - ct) + f(x + ct)] = \sum_{m=1}^{\infty} u_m \sin\left(\frac{m\pi x}{l}\right) \cos\left(\frac{m\pi ct}{l}\right),$$

$$\frac{1}{2c} \int_{x-ct}^{x+ct} g(s)ds = \frac{l}{\pi c} \sum_{m=1}^{\infty} \frac{v_m}{m} \sin\left(\frac{m\pi x}{l}\right) \sin\left(\frac{m\pi ct}{l}\right).$$

The solution $y(x, t)$ becomes

$$y(x, t) = \sum_{m=1}^{\infty} u_m \sin\left(\frac{m\pi x}{l}\right) \cos\left(\frac{m\pi ct}{l}\right)$$

$$+ \frac{l}{\pi c} \sum_{m=1}^{\infty} \frac{v_m}{m} \sin\left(\frac{m\pi x}{l}\right) \sin\left(\frac{m\pi ct}{l}\right). \tag{2.3.4}$$

Check that $y(x, t)$ given in (2.3.4) actually satisfies the initial and the boundary conditions.

**Note**: In order that $y(x, t)$ given in (2.3.4) is a solution of the given p.d.e., the series on the right-hand side of (2.3.4) must converge. In addition, the series obtained by term-by-term differentiation twice with respect to $x$ and $t$ should also converge. A sufficient condition for this is that the function $u(x)$ is twice continuously differentiable and its third derivative is piecewise continuous (i.e., the number of points of discontinuity of $u'''(x)$ is finite in the interval $(0, l)$, and at each point of discontinuity the left and right hand limits exist). In addition, $v(x)$ is continuously differentiable and its second derivative is piecewise continuous (refer to Miroslaw Krżyzański).

**Exercise 2.3.1**: Let $u(x, t)$ be any solution of Equation (2.3.1) and $A$, $B$, $C$, and $D$ be the vertices of any parallelogram whose sides are characteristic curves (i.e., the lines parallel to $x + ct = 0$ and $x - ct = 0$). Then show that

$$u(A) + u(C) = u(B) + u(D).$$

(Hint: Make use of the general solution of Equation (2.3.1)). Make use of the previous result to find $u(\frac{1}{2}, \frac{3}{2})$ if $u$ satisfies

$$u_{tt} = u_{xx}, \quad 0 < x < 1, \ t > 0,$$

$$u(x, 0) = u_t(x, 0) = 0,$$

$$u(0, t) = \sin \pi t, \ u(1, t) = t.$$

## 2.3.4   Riemann's Method

Consider

$$L[u] = u_{xy} + a(x, y)u_x + b(x, y)u_y + c(x, y)u = f(x, y), \tag{2.3.5}$$

where $a$, $b$, $c$, and $f$ are continuously differentiable functions with respect to $x$ and $y$. This is a linear, second order hyperbolic equation that is in the canonical form. The characteristics are $x = $ constant and $y = $ constant.

Let $v(x)$ be a function having continuous second order partial derivatives. Then

$$vu_{xy} - uv_{xy} = (vu_x)_y - (uv_y)_x ,$$
$$avu_x = (avu)_x - u(av)_x ,$$
$$bvu_y = (bvu)_y - u(bv)_y ,$$

so that

$$vL[u] - uM[v] = U_x + V_y , \tag{2.3.6}$$

where

$$M[v] = v_{xy} - (av)_x - (bv)_y + cv , \tag{2.3.7}$$
$$U = auv - uv_y , \ \text{and}$$
$$V = buv + vu_x .$$

The operator $M$ is called the adjoint operator of $L$. (If $M = L$, then the operator $L$ is said to be self-adjoint.)

We will now state a theorem without proof that we shall use again in Section 2.4.

**Theorem 2.3.1**: **(Green's theorem)**

Let $C$ be a closed curve bounding the region of integration $D$ and $U$ and $V$ be differentiable functions in $D$ and continuous on $C$. Then Green's theorem states that

$$\int\int_D (U_x + V_y)dxdy = \oint_C (Udy - Vdx). \qquad \square$$

We will now discuss the Cauchy problem (refer to Section 2.4.3) for the hyperbolic equation (2.3.5). Let $\Gamma$ be a smooth initial curve. We assume that the tangent to $\Gamma$ is nowhere parallel to the $x$ or $y$ axes (as $x = $ constant, $y = $ constant are characteristics of (2.3.5)). We suppose that $u$ and $u_x$ (or $u_y$) are prescribed along $\Gamma$. We want to find the solution of (2.3.5) in the neighborhood of $\Gamma$.

Let $P(\alpha, \beta)$ be a point at which the solution to the Cauchy problem is sought. Let the characteristics through $P$ intersect the initial curve $\Gamma$ at $Q$ and $R$ (refer to Fig. 2.3.4).

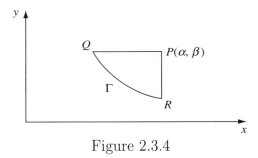

Figure 2.3.4

Let $D$ be the region bounded by the closed contour $PQRP$ (say, $C$). We shall now apply Green's theorem to this region. On using the identity (2.3.6), we obtain

$$\int\int_D (vLu - uMv)dxdy = \int_Q^R (Udy - Vdx) + \int_R^P Udy - \int_P^Q Vdx.$$

Now let us compute the last term on the right-hand side of the previous equation

$$\int_P^Q Vdx = \int_P^Q buvdx + \int_P^Q vu_xdx,$$

$$= [uv]_P^Q + \int_P^Q u(bv - v_x)dx.$$

Substituting this back into the equation gives

$$[uv]_P = [uv]_Q + \int_P^Q u(bv - v_x)dx + \int_P^R u(av - v_y)dy$$

$$- \int_Q^R (Udy - Vdx) + \int\int_D (vLu - uMv)dxdy.$$

Let us, if possible, choose the function $v(x, y; \alpha, \beta)$ to be the solution of the adjoint equation $M[v] = 0$, satisfying the conditions

$$v_x = bv \quad \text{on} \quad y = \beta,$$

$$v_y = av \quad \text{on} \quad x = \alpha,$$

$$v = 1 \quad \text{at} \quad x = \alpha \quad \text{and} \quad y = \beta.$$

Such a function $v(x, y; \alpha, \beta)$, if it exists, is called a **Riemann function**.
**Note**: The Riemann function is the solution of a hyperbolic equation where the data is prescribed on both the characteristics passing through the point $P(\alpha, \beta)$. Since $L[u] = f$, we obtain

$$[u]_P = [uv]_Q - \int_Q^R uv(ady - bdx)$$

$$+ \int_Q^R (uv_y dy + vu_x dx) + \int\int_D vf dxdy. \tag{2.3.8}$$

Equation (2.3.8) gives $u$ at $P$ when $u$ and $u_x$ are prescribed along the curve $\Gamma$. However, if $u$ and $u_y$ are prescribed, the identity

$$[uv]_R - [uv]_Q = \int_Q^R [(uv)_x dx + (uv)_y dy],$$

gives, along with (2.3.8),

$$[u]_P = [uv]_R - \int_Q^R uv(ady - bdx) - \int_Q^R (uv_x dx + vu_y dy)$$

$$+ \int\int_D vf dxdy. \tag{2.3.9}$$

Further, on adding (2.3.8) and (2.3.9), we get

$$[u]_P = \frac{1}{2}([uv]_Q + [uv]_R) - \int_Q^R uv(a\,dy - b\,dx)$$

$$- \frac{1}{2}\int_Q^R u(v_x dx - v_y dy) + \frac{1}{2}\int_Q^R v(u_x dx - u_y dy)$$

$$+ \int\int_D vf\,dxdy. \tag{2.3.10}$$

**Note**: The solution at the point $P(\alpha, \beta)$ depends only on the Cauchy data along the arc $QR$ on $\Gamma$.

**Example 2.3.1**: Prove that for the equation

$$Lu = u_{xy} + \frac{1}{4}u = 0,$$

the Riemann function is

$$v(x, y; \alpha, \beta) = J_0(\sqrt{(x - \alpha)(y - \beta)}\,),$$

where $J_0$ denotes the Bessel's function of the first kind of order zero.
**Solution**: The Riemann function $v$ is the solution of the adjoint equation

$$M[v] = v_{xy} + \frac{1}{4}v = 0, \tag{2.3.11}$$

which satisfies the following conditions

$$v_x = 0 \ \text{ on } y = \beta,$$

$$v_y = 0 \ \text{ on } x = \alpha,$$

$$v = 1 \ \text{ at } \ x = \alpha \ \text{ and } \ y = \beta.$$

Let $\eta = (x - \alpha)(y - \beta)$ and $v(x, y; \alpha, \beta) = z(\eta)$.
Then

$$v_x = z_\eta(y - \beta),$$
$$v_y = z_\eta(x - \alpha),$$
$$v_{xy} = z_{\eta\eta}(x - \alpha)(y - \beta) + z_\eta.$$

Therefore $v_x = 0$ on $y = \beta$, $v_y = 0$ on $x = \alpha$.
Equation (2.3.11) is transformed to the ordinary differential equation

$$\eta z_{\eta\eta} + z_\eta + \frac{1}{4}z = 0,$$

where $z(\eta)$ satisfies the condition

$$z(0) = 1.$$

Therefore $z(\eta) = J_0(\sqrt{\eta})$. □

**Example 2.3.2**: For the wave equation in canonical form

$$u_{\xi\eta} = 0, \qquad\qquad (2.3.12)$$

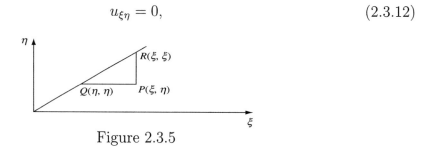

Figure 2.3.5

the Riemann function is $v(\xi, \eta; \alpha, \beta) \equiv 1$ and hence the solution is given by

$$u[P] = \frac{1}{2}[u(Q) + u(R)] + \frac{1}{2}\int_{QR}(u_\xi d\xi - u_\eta d\eta).$$

Introducing new coordinates

$$x = (\xi + \eta)/2,$$
$$t = (\xi - \eta)/2c,$$

Equation (2.3.12) becomes

$$\frac{1}{c^2}u_{tt} - u_{xx} = 0,$$

and

$$u[P] = \frac{1}{2}[u(Q) + u(R)] + \frac{1}{2}\int_Q^R \left(u_x + \frac{1}{c}u_t\right)\frac{1}{2}(dx + cdt)$$

$$- \frac{1}{2}\int_Q^R \left(u_x - \frac{1}{c}u_t\right)\frac{1}{2}(dx - cdt),$$

$$= \frac{1}{2}[u(Q) + u(R)] + \frac{1}{2}\int_Q^R \left(\frac{1}{c}u_t dx + cu_x dt\right).$$

When the initial curve is the line $t = 0$, i.e., $\xi = \eta$, then $Q$ and $R$ have the coordinates $(x - ct, 0)$, $(x + ct, 0)$ .

Therefore

$$u[P] = \frac{1}{2}[u(x - ct, 0) + u(x + ct, 0)] + \frac{1}{2c} \int_Q^R u_t dx, \qquad (2.3.13)$$

which is **d'Alembert's solution**.     □

**Exercise 2.3.2**: Show that

$$v(x, y; \alpha, \beta) = \frac{(x + y)[2xy + (\alpha - \beta)(x - y) + 2\alpha\beta]}{(\alpha + \beta)^3} ,$$

is the Riemann function for the second order p.d.e.

$$u_{xy} + \frac{2}{x + y}(u_x + u_y) = 0.$$

## 2.3.5   Vibrations of a String of Finite Length (Method of Separation of Variables)

Let us consider the following problem

$$y_{tt} - c^2 y_{xx} = 0, \qquad 0 < x < l, \ t > 0, \qquad (2.3.14)$$
$$y(x, 0) = f(x), \qquad 0 \le x \le l, \qquad (2.3.15)$$
$$y_t(x, 0) = g(x), \qquad 0 \le x \le l, \qquad (2.3.16)$$
$$y(0, t) = y(l, t) = 0, \qquad t > 0. \qquad (2.3.17)$$

So $f$ and $g$ are the initial displacement and velocity respectively. Let us assume the solution of Equation (2.3.14) in the form

$$y(x, t) = X(x)T(t).$$

Then

$$\frac{X''}{X} = \frac{T''}{c^2 T} .$$

Observe that the right-hand side is a function of $t$ alone while the left-hand side is a function of $x$ alone. Hence each of them must be constant and equal to, say, $\lambda$. Therefore

$$X'' - \lambda X = 0,$$
$$T'' - c^2 \lambda T = 0.$$

From (2.3.17) we have

$$y(0,t) = X(0)T(t) = 0 \ \forall \ t \geq 0.$$

Since $T(t) \not\equiv 0$, we get $X(0) = 0$.
Similarly $y(l,t) = 0 \Rightarrow X(l) = 0$.
Therefore we have

$$X'' = \lambda X,$$

$$X(0) = 0 = X(l),$$

which is an eigenvalue problem.

**Case (i):** $\lambda > 0$. The solution of the previous eigenvalue problem is

$$X(x) = Ae^{-\sqrt{\lambda}x} + Be^{\sqrt{\lambda}x},$$

where $A$ and $B$ are arbitrary constants. To satisfy the boundary conditions

$$A + B = Ae^{-\sqrt{\lambda}l} + Be^{\sqrt{\lambda}l} = 0,$$

the only possibility is $A = B = 0$. Hence there is no eigenvalue $\lambda > 0$.
**Case (ii):** $\lambda = 0$. In this case the solution of the eigenvalue problem is of the form

$$X(x) = A + Bx.$$

The boundary conditions imply that $A = 0$ and $A + Bl = 0$.
Therefore $A = B = 0$. Hence $\lambda = 0$ is not an eigenvalue.
**Case (iii):** $\lambda < 0$. The solution in this case is of the form

$$X(x) = A \cos \sqrt{-\lambda}x + B \sin \sqrt{-\lambda}x.$$

The condition $X(0) = 0$ implies that $A = 0$ and $X(l) = 0$ implies that $B \sin \sqrt{-\lambda}l = 0$.
As $B = 0$ gives only a trivial solution, we must have $\sin \sqrt{-\lambda}l = 0$ for a non-trivial solution. Therefore

$$\sqrt{-\lambda}l = n\pi, n = 1, 2, 3, \cdots,$$

$$-\lambda_n = \left(\frac{n\pi}{l}\right)^2.$$

These $\lambda_n$ are called eigenvalues and the functions $\sin(n\pi x/l)$ are the corresponding eigenfunctions. Therefore $X_n = B_n \sin(n\pi x/l)$.

For each $\lambda_n$, we have

$$T_n(t) = C_n \cos\left(\frac{n\pi ct}{l}\right) + D_n \sin\left(\frac{n\pi ct}{l}\right),$$

where $C_n$ and $D_n$ are arbitrary constants. Hence

$$y_n(x,t) = \left(a_n \cos\left(\frac{n\pi ct}{l}\right) + b_n \sin\left(\frac{n\pi ct}{l}\right)\right) \sin\left(\frac{n\pi x}{l}\right),$$

is a solution of Equation (2.3.14) and satisfies the boundary conditions (2.3.17).

If $y_1$ and $y_2$ are two solutions of a linear, homogeneous equation satisfying linear, homogeneous boundary conditions, then $y_1 + y_2$ is also a solution of that equation and satisfies the same boundary conditions. This is called the principle of superposition. Observe that Equation (2.3.14) and the boundary conditions (2.3.17) are linear and homogeneous. Therefore, by the principle of superposition, the series

$$y(x,t) = \sum_{n=1}^{\infty} y_n(x,t), \tag{2.3.18}$$

if it converges, is also a solution of Equation (2.3.14) satisfying the boundary conditions (2.3.17). In fact, we assume that term-by-term differentiation is possible and that the derived series is also convergent. Now $a_n$ and $b_n$ must be chosen such that $y$ as given in Equation (2.3.18) satisfies the initial conditions (2.3.15) and (2.3.16). The initial condition $y(x,0) = f(x)$ gives

$$f(x) = \sum_{n=1}^{\infty} a_n \sin\left(\frac{n\pi x}{l}\right), \quad 0 < x < l. \tag{2.3.19}$$

Similarly the initial condition $y_t(x,0) = g(x)$ gives

$$g(x) = \sum_{n=1}^{\infty} b_n \left(\frac{n\pi c}{l}\right) \sin\left(\frac{n\pi x}{l}\right), \quad 0 < x < l. \tag{2.3.20}$$

Hence $a_n$ and $b_n$ are given by the Fourier coefficients of the half range sine series of $f(x)$ and $g(x)$ respectively. Therefore

$$a_n = \frac{2}{l} \int_0^l f(x) \sin\left(\frac{n\pi x}{l}\right) dx, \tag{2.3.21}$$

and

$$b_n = \frac{2}{n\pi c} \int_0^l g(x) \sin\left(\frac{n\pi x}{l}\right) dx. \qquad (2.3.22)$$

We derived the same result earlier in Section 2.3.3 from d'Alembert's solution.

**Exercise 2.3.4**: Solve

$$y_{tt} - c^2 y_{xx} = 0, \quad 0 < x < 1, \quad t > 0,$$
$$y(0, t) = y(1, t) = 0,$$
$$\text{(i)} \quad y(x, 0) = x(1 - x), \quad 0 \le x \le 1,$$
$$y_t(x, 0) = 0, \quad 0 \le x \le 1.$$
$$\text{(ii)} \quad y(x, 0) = 0, \quad 0 \le x \le 1,$$
$$y_t(x, 0) = x^2, \quad 0 \le x \le 1.$$

**Theorem 2.3.2: Uniqueness of solution.**

The solution of the following problem, if it exists, is unique.

$$u_{tt} - c^2 u_{xx} = F(x, t), \quad 0 < x < l, \quad t > 0,$$
$$u(x, 0) = f(x), \quad 0 \le x \le l,$$
$$u_t(x, 0) = g(x), \quad 0 \le x \le l,$$
$$u(0, t) = u(l, t) = 0, \quad t \ge 0.$$

**Proof**: Let there be two solutions $u_1$ and $u_2$ of the previous problem. Then $v = (u_1 - u_2)$ will be a solution of the following problem.

$$v_{tt} - c^2 v_{xx} = 0, \quad 0 < x < l, \quad t > 0,$$
$$v(x, 0) = 0, \quad 0 \le x \le l,$$
$$v_t(x, 0) = 0, \quad 0 \le x \le l,$$
$$v(0, t) = v(l, t) = 0, \quad t \ge 0.$$

We shall show that $v \equiv 0$, which would imply that $u_1 \equiv u_2$.
Consider

$$E(t) = \frac{1}{2} \int_0^l (c^2 v_x^2 + v_t^2) dx.$$

Observe that $E(t)$ is a differentiable function of $t$, since $v(x, t)$ is twice differentiable. Therefore

$$\frac{dE}{dt} = \int_0^l (c^2 v_x v_{xt} + v_t v_{tt}) dx,$$
$$= \int_0^l v_t v_{tt} dx + [c^2 v_x v_t]_0^l - \int_0^l c^2 v_t v_{xx} dx.$$

$$v(0, t) = 0 \Rightarrow v_t(0, t) = 0 \ \forall \ t \geq 0,$$

$$v(l, t) = 0 \Rightarrow v_t(l, t) = 0 \ \forall \ t \geq 0.$$

Therefore

$$\frac{dE}{dt} = \int_0^l v_t(v_{tt} - c^2 v_{xx})dx = 0.$$

Therefore $E = $ constant.
Given that $v_t(x, 0) = 0$ and since $v(x, 0) = 0$ implies $v_x(x, 0) = 0$, we obtain

$$E(0) = 0.$$

Therefore $E \equiv 0$.
Hence $v_x \equiv 0, \ v_t \equiv 0 \ \forall \ t > 0, \ 0 < x < l$.
This is possible only if $v(x, t) = $ constant.
Since $v(x, 0) = 0, \ v \equiv 0$.
Hence the theorem. $\qquad\qquad\qquad\qquad\qquad\qquad\qquad\qquad\qquad$ $\square$

**Note**: The solution of the problem of vibrations of a string of finite length, stated in Equations (2.3.14)–(2.3.17) is also unique, as it is a special case of Theorem 2.3.2, i.e., the case when $F(x, t) = 0$.

## 2.4   Laplace's Equation

The Laplace's equation in two dimensions is,

$$\nabla^2 u = \frac{\partial^2 u}{\partial x^2} + \frac{\partial^2 u}{\partial y^2} \equiv 0. \qquad\qquad (2.4.1)$$

A solution of (2.4.1) is called a two-dimensional harmonic function.

### 2.4.1   Boundary Value Problems

Let $D$ be the interior of a simple, closed, smooth curve $B$ and $f$ be a continuous function defined on the boundary $B$.

- **The first boundary value problem:  The Dirichlet problem.**
  The problem of finding a harmonic function $u(x, y)$ in $D$ such that it coincides with $f$ on the boundary $B$ is called the Dirichlet problem.

- **The second boundary value problem: The Neumann problem.**
  This involves finding a function $u(x, y)$ such that it is harmonic in $D$ and satisfies $\dfrac{\partial u}{\partial n} = f(s)$ on $B$, (where $\dfrac{\partial}{\partial n}$ is the directional derivative along the outward normal) with the condition $\int_B f(s)ds = 0$. In fact, the vanishing of the integral is a necessary condition for the solution to exist (refer to Theorem 2.4.4).

- **The third boundary value problem: The Robin Problem.**
  This involves finding a function $u(x, y)$ that is harmonic in $D$ and satisfies the condition $\dfrac{\partial u}{\partial n} + h(s)u(s) = 0$ on $B$ where $h(s) \geq 0$, and $h(s) \not\equiv 0$.

- **The fourth boundary value problem:**
  This involves finding a function $u(x, y)$ that is harmonic in $D$ and satisfies the boundary conditions of different types on different portions of the boundary $B$. For example, $u = f_1(s)$ on $B_1$, $\dfrac{\partial u}{\partial n} = f_2(s)$ on $B_2$, where $B_1 \cup B_2 = B$.

## 2.4.2   Maximum and Minimum Principles

**Theorem 2.4.1:**   Suppose that $u(x, y)$ is harmonic in a bounded domain $D$ and continuous in $\bar{D} = D \cup B$. Then $u$ attains its maximum on the boundary $B$ of $D$.

**Proof**: Let the maximum of $u$ on $B$ be $M$. Let us now suppose that the maximum of $u$ on $\bar{D}$ is not attained at any point on $B$. Then it must be attained at some point $P(x_0, y_0)$ in $D$.

If $M_0 = u(x_0, y_0)$, then $M_0 > M$.

Consider

$$v(x, y) = u(x, y) + \frac{M_0 - M}{4R^2}[(x - x_0)^2 + (y - y_0)^2],$$

where $(x, y) \in D$ and $R$ is the radius of the circle with center $(x_0, y_0)$ containing $D$. Such an $R$ exists since $D$ is bounded. Observe that $v(x_0, y_0) = u(x_0, y_0) = M_0$. On $B$, we have

$$v(x, y) \leq M + \frac{(M_0 - M)}{4} < M_0.$$

Thus $v(x, y)$, like $u(x, y)$, must attain its maximum at a point in $D$.

Therefore $v_{xx} \leq 0, v_{yy} \leq 0$ at some point in $D$, i.e., $v_{xx} + v_{yy} \leq 0$ at some point in $D$.

However, in $D$

$$v_{xx} + v_{yy} = u_{xx} + u_{yy} + \frac{(M_0 - M)}{R^2} = \frac{(M_0 - M)}{R^2} > 0,$$

which is a contradiction. Hence the maximum of $u$ must be attained on $B$. □

**Theorem 2.4.2**: Suppose that $u(x, y)$ is harmonic in a bounded domain $D$ and is continuous on $\bar{D} = D \cup B$. Then $u$ attains its minimum on the boundary $B$ of $D$.

**Proof**: The proof is immediate on applying Theorem 2.4.1 to the harmonic function $-u(x, y)$. □

**Theorem 2.4.3**: **Uniqueness theorem**

The solution of the Dirichlet problem, if it exists, is unique.

**Proof**: Suppose $u_1(x, y)$ and $u_2(x, y)$ are two solutions of the Dirichlet problem. That is

$$\nabla^2 u_1 = 0 \quad \text{in} \quad D \quad \text{and} \quad u_1 = f(s) \quad \text{on} \quad B,$$

$$\nabla^2 u_2 = 0 \quad \text{in} \quad D \quad \text{and} \quad u_2 = f(s) \quad \text{on} \quad B.$$

As $u_1$ and $u_2$ are harmonic in $D$, $(u_1 - u_2)$ is also harmonic in $D$. However $(u_1 - u_2) = 0$ on $B$. By the maximum and minimum principle, $(u_1 - u_2) \equiv 0$ in $D$. Hence the theorem. □

**Green's identities**:

If $U(x, y)$ and $V(x, y)$ are functions defined inside and on the boundary $B$ of the closed region $D$ (refer to Green's theorem on page 95 ), then

$$\int_D \left( \frac{\partial U}{\partial x} + \frac{\partial V}{\partial y} \right) dS = \int_B (U \, dy - V \, dx).$$

Let

$$U = \psi \frac{\partial \phi}{\partial x}, \quad V = \psi \frac{\partial \phi}{\partial y}.$$

Then

$$\int_D (\psi_x \phi_x + \psi \phi_{xx} + \psi_y \phi_y + \psi \phi_{yy}) dS = \int_B \psi \frac{\partial \phi}{\partial n} ds. \qquad (2.4.2)$$

On interchanging $\phi$ and $\psi$ and subtracting one from the other, we get

$$\int_D (\psi \nabla^2 \phi - \phi \nabla^2 \psi) dS = \int_B (\psi \frac{\partial \phi}{\partial n} - \phi \frac{\partial \psi}{\partial n}) ds. \qquad (2.4.3)$$

The identities (2.4.2) and (2.4.3) are called Green's identities.    □.

**Theorem 2.4.4: The necessary condition for the Neumann problem**

Let $u$ be a solution of the Neumann problem

$$\nabla^2 u = 0 \quad \text{in} \ D,$$

and

$$\frac{\partial u}{\partial n} = f(s) \quad \text{on} \ B.$$

Then

$$\int_B f(s)ds = 0. \tag{2.4.4}$$

**Proof**: Put $\psi = 1$ and $\phi = u$ in (2.4.3). Then we get

$$\int_B f(s)ds = 0.$$

Thus $f(s)$ cannot be arbitrarily prescribed as it has to satisfy the previous condition.
□

**Theorem 2.4.5**: The solution of the Neumann problem is unique up to the addition of a constant.

**Proof**: Let $u_1$ and $u_2$ be two harmonic functions in $D$ bounded by $B$ such that

$$\frac{\partial u_1}{\partial n} = \frac{\partial u_2}{\partial n} = f(s) \quad \text{on} \ B.$$

Then we will show that $(u_1 - u_2) = \text{constant}$. Consider

$$v = u_1 - u_2.$$

Then

$$\nabla^2 v = 0,$$

and

$$\frac{\partial v}{\partial n} = 0 \quad \text{on} \ B.$$

Take $\phi = \psi = v$ in (2.4.2). We get

$$\int_D (\nabla v)^2 dS = 0.$$

This implies that $\nabla v = 0$ on $D$ since $\nabla v$ is a continuous function. So $v = \text{constant}$.
□

## 2.4.3   The Cauchy Problem

Let

$$Au_{xx} + Bu_{xy} + Cu_{yy} = F(x, y, u, u_x, u_y), \qquad (2.4.5)$$

where $A(x, y), B(x, y)$, and $C(x, y)$, are functions of $x$ and $y$ and $\Gamma_0$ be a curve in the $x, y$-plane. The problem of finding the solution $u(x, y)$ of the p.d.e. (2.4.5) in the neighborhood of $\Gamma_0$ satisfying the following conditions

$$u = f(s), \qquad (2.4.6)$$

$$\frac{\partial u}{\partial n} = g(s), \qquad (2.4.7)$$

on $\Gamma_0(s)$ is called a Cauchy problem. The conditions (2.4.6) and (2.4.7) are called the Cauchy conditions.

**Example**: The problem of vibrations of an infinite string solved earlier is a Cauchy problem. □

**Stability**: Data in nature cannot possibly be conceived as rigidly fixed; the mere process of measuring involves small errors. Therefore a mathematical problem cannot be formulated realistically corresponding to physical phenomena unless a slight variation of the given initial and/or boundary data leads to only a slight variation in the solution. In such cases, we say that the solution depends continuously on the initial and/or boundary data. Such a solution is said to be stable.

Consider the Cauchy problem in the case of Laplace's equation

$$u_{xx} + u_{yy} = 0,$$

with the following data prescribed on the $x$-axis,

$$u(x, 0) = 0,$$

$$u_y(x, 0) = n^{-1} \sin nx.$$

The initial curve here is the $x$-axis. It is easy to verify that

$$u(x, y) = n^{-2} \sinh ny \sin nx,$$

is the solution of the previous problem. Observe that when $n \to \infty$ the function $n^{-1} \sin nx \to 0$ uniformly. However $n^{-2} \sinh ny \sin nx$ does not become small as $n \to \infty$ for any $y \neq 0$. Therefore the solution is not stable. This example is due to Hadamard.

We shall now show that the solution to the Dirichlet problem is stable, i.e., it depends continuously on the boundary data.

Let $u_1$ and $u_2$ be the solutions of

$$\nabla^2 u_1 = 0 \ \text{ in } \ D,$$

$$u_1 = f_1 \ \text{ on } \ B,$$

and

$$\nabla^2 u_2 = 0 \ \text{ in } \ D,$$

$$u_2 = f_2 \ \text{ on } \ B.$$

Let $v = (u_1 - u_2)$. Then

$$\nabla^2 v = 0 \ \text{ in } \ D,$$

$$v = f_1 - f_2 \ \text{ on } \ B.$$

By the maximum and minimum principle, the harmonic function $v$ attains its maximum and minimum on $B$, which is nothing but the maximum and minimum of $f_1 - f_2$ on $B$. Thus if $\mid f_1 - f_2 \mid < \epsilon$ on $B$, then $-\epsilon \leq v_{\min} \leq v_{\max} \leq \epsilon$ on $D$, i.e., $\mid v \mid \leq \epsilon$ on $D$. Hence $\mid u_1 - u_2 \mid < \epsilon$. Thus the solution depends continuously on the boundary data. Note that d'Alembert's solution of wave equation is stable.

**Note:** For any physical problem to be meaningful, we require the existence, uniqueness, and stability of the solution. These three conditions are called Hadamard's conditions for a **well-posed problem**.

## 2.4.4   The Dirichlet Problem for the Upper Half Plane

Consider the following problem

$$u_{xx} + u_{yy} = 0, \ -\infty < x < \infty, \ y > 0, \tag{2.4.8}$$

$$u(x,0) = f(x), \ -\infty < x < \infty, \tag{2.4.9}$$

with the conditions that $u$ is bounded as $y \to \infty$, $u$ and $u_x$ vanish as $\mid x \mid \to \infty$. Observe that this is a Dirichlet problem for the upper half plane.

Suppose $U(\alpha, y)$ is the Fourier transform of $u(x, y)$ in the variable $x$. Then by definition (refer to Appendix A)

$$U(\alpha, y) = \frac{1}{\sqrt{2\pi}} \int_{-\infty}^{\infty} u(x, y) e^{i\alpha x} dx.$$

Then by applying the Fourier transform to (2.4.8), it becomes

$$U_{yy} - \alpha^2 U = 0,$$

whose solution is of the form

$$U = A(\alpha)e^{\alpha y} + B(\alpha)e^{-\alpha y} .$$

Since we require that $u$ be bounded as $y \to \infty, U(\alpha, y)$ must also be bounded as $y \to \infty$.
Hence for $\alpha > 0$,   $A(\alpha) = 0$ and for $\alpha < 0$,   $B(\alpha) = 0$. Therefore

$$U(\alpha, y) = U(\alpha, 0)e^{-|\alpha|y},$$

$$U(\alpha, 0) = \mathcal{F}[u(x, 0)] = \mathcal{F}[f(x)] = F(\alpha).$$

Therefore

$$U(\alpha, y) = F(\alpha)e^{-|\alpha|y}.$$

We have

$$\mathcal{F}^{-1}(e^{-|\alpha|y}) = \sqrt{\frac{2}{\pi}} \left( \frac{y}{y^2 + x^2} \right).$$

Therefore by the Convolution theorem (refer to Theorem A2 in Appendix A)

$$u(x, y) = f(x) * \sqrt{\frac{2}{\pi}} \left( \frac{y}{y^2 + x^2} \right),$$

$$= \frac{1}{\sqrt{2\pi}} \int_{-\infty}^{\infty} f(\xi) \sqrt{\frac{2}{\pi}} \left( \frac{y}{y^2 + (x - \xi)^2} \right) d\xi,$$

$$= \frac{y}{\pi} \int_{-\infty}^{\infty} \frac{f(\xi)}{y^2 + (x - \xi)^2} d\xi. \tag{2.4.10}$$

## 2.4.5   The Neumann Problem for the Upper Half Plane

Consider the following problem

$$u_{xx} + u_{yy} = 0, \quad -\infty < x < \infty, \ y > 0, \tag{2.4.11}$$

$$u_y(x, 0) = g(x), \quad -\infty < x < \infty, \tag{2.4.12}$$

with the conditions that $u$ is bounded as $y \to \infty$, $u$ and $u_x$ vanish as $| x | \to \infty$ and

$$\int_{-\infty}^{\infty} g(x)dx = 0,$$

which is the necessary condition for the solution to exist. We now find the solution to this problem by introducing a new variable $v(x, y)$ as $v(x, y) = u_y(x, y)$. Then

$$u = \int_{a}^{y} v(x, \eta)d\eta.$$

The problem when reformulated in terms of $v$ is stated as follows

$$\nabla^2 v(x, y) = \nabla^2 u_y = \frac{\partial}{\partial y}(\nabla^2 u) = 0,$$

$$v(x, 0) = u_y(x, 0) = g(x).$$

Therefore, from Equation (2.4.10), we get

$$v(x, y) = \frac{y}{\pi} \int_{-\infty}^{\infty} \frac{g(\xi)}{(\xi - x)^2 + y^2} d\xi.$$

Hence

$$u(x, y) = \frac{1}{\pi} \int_{a}^{y} \eta \int_{-\infty}^{\infty} \frac{g(\xi)}{(\xi - x)^2 + \eta^2} d\xi d\eta,$$

$$= \frac{1}{2\pi} \int_{-\infty}^{\infty} g(\xi) \log \left[ \frac{(\xi - x)^2 + y^2}{(\xi - x)^2 + a^2} \right] d\xi. \tag{2.4.13}$$

## 2.4.6   The Dirichlet Problem for a Circle

We now find the solution of the Dirichlet problem for a circle of radius $a$. The problem is to solve for $u$ from

$$\nabla^2 u = u_{rr} + \frac{1}{r} u_r + \frac{1}{r^2} u_{\theta\theta} = 0, \quad r < a, \tag{2.4.14}$$

subject to the boundary condition

$$u(a, \theta) = f(\theta). \tag{2.4.15}$$

As Equation (2.4.14) is linear and homogeneous, we assume a variable separable solution of the form

$$u(r, \theta) = R(r)H(\theta).$$

Then Equation (2.4.14) yields

$$\frac{r^2 R''}{R} + r\frac{R'}{R} = -\frac{H''}{H} = \lambda,$$

where $\lambda$ is a constant. This implies that

$$r^2 R'' + rR' - \lambda R = 0,$$

$$H'' + \lambda H = 0.$$

$\lambda < 0$ does not give any acceptable solution, since $H(\theta + 2\pi) = H(\theta)$ is not met in this case.

$\lambda - 0$ gives the solution in the form

$$u(r, \theta) = (A + B \log r)(C\theta + D).$$

The periodicity of $H$ gives $C = 0$. $r = 0$ is a point in the domain and since $u$ must be bounded there, we must have $B = 0$. Therefore, in this case, $u = $ constant. Let $\lambda > 0$. Assume $\lambda = \alpha^2$. Then

$$H(\theta) = A \cos \alpha\theta + B \sin \alpha\theta.$$

The periodicity condition implies $\alpha = 1, 2, 3 \cdots$. Then

$$R(r) = Cr^\alpha + Dr^{-\alpha}.$$

Since $r^{-\alpha} \to \infty$ as $r \to 0, D$ must be zero. Thus the general solution of Equation (2.4.14) is

$$u(r, \theta) = \frac{a_0}{2} + \sum_{n=1}^{\infty} \left(\frac{r}{a}\right)^n (a_n \cos n\theta + b_n \sin n\theta), \tag{2.4.16}$$

where $a_0, a_n$, and $b_n$ are constants. Then the boundary conditions $u(a, \theta) = f(\theta)$ gives

$$a_n = \frac{1}{\pi} \int_0^{2\pi} f(\theta) \cos n\theta \ d\theta, \quad n = 0, 1, 2, \cdots,$$

and

$$b_n = \frac{1}{\pi} \int_0^{2\pi} f(\theta) \sin n\theta \ d\theta, \quad n = 1, 2, \cdots.$$

We will now show that the previous series given in Equation (2.4.16) converges.

By the very definition, $a_n$ and $b_n$ are bounded. Choose $M$ such that $\mid a_n \mid < M$ and $\mid b_n \mid < M, n = 1, 2, \cdots$. Since

$$u_n(\rho, \theta) = \rho^n(a_n \cos n\theta + b_n \sin n\theta), \quad \rho = \frac{r}{a},$$

we have

$$\mid u_n \mid \quad < 2\rho_0^n M, \quad 0 \le \rho \le \rho_0 < 1.$$

Hence, in any closed circular region inside the open unit disc, the series converges uniformly.

Also observe

$$\mid \frac{\partial u_n}{\partial r} \mid = \mid \frac{n}{a} \rho^{n-1}(a_n \cos n\theta + b_n \sin n\theta) \mid,$$

$$< \frac{2n}{a} \rho_0^{n-1} M.$$

Thus the series obtained by differentiating (2.4.16) term by term with respect to $r$ converges uniformly in any closed circular region inside the open unit disc.

In a similar manner, we can prove that the series obtained by twice differentiating the series (2.4.16) term by term with respect to $r$ and $\theta$ converge uniformly in any closed circular region inside the open unit disc. Consequently,

$$\nabla^2 u = u_{rr} + \frac{1}{r} u_r + \frac{1}{r^2} u_{\theta\theta},$$

$$= \sum_{n=2}^{\infty} \frac{\rho^{n-2}}{a^2}(a_n \cos n\theta + b_n \sin n\theta)[n(n-1) + n - n^2],$$

$$= 0, \quad 0 \le \rho \le \rho_0 < 1.$$

Thus $u$ is harmonic in the region $0 \le \rho < 1$.

Consider

$$u(\rho, \theta) = \frac{1}{2\pi} \int_0^{2\pi} f(\tau)d\tau$$

$$+ \frac{1}{\pi} \sum_{n=1}^{\infty} \rho^n \int_0^{2\pi} f(\tau)[\cos n\tau \cos n\theta + \sin n\tau \sin n\theta]d\tau,$$

$$= \frac{1}{2\pi} \int_0^{2\pi} [1 + 2 \sum_{n=1}^{\infty} \rho^n \cos n(\theta - \tau)]f(\tau)d\tau.$$

Here, due to the uniform convergence of the series, the interchange of summation and integration is allowed.

For $0 \leq \rho < 1$,

$$[1 + 2\sum_{n=1}^{\infty} \rho^n \cos n(\theta - \tau)] = 1 + \sum_{n=1}^{\infty} [\rho^n e^{in(\theta - \tau)} + \rho^n e^{-in(\theta - \tau)}],$$

$$= 1 + \frac{\rho e^{i(\theta - \tau)}}{1 - \rho e^{i(\theta - \tau)}} + \frac{\rho e^{-i(\theta - \tau)}}{1 - \rho e^{-i(\theta - \tau)}},$$

$$= \frac{1 - \rho^2}{1 - \rho e^{i(\theta - \tau)} - \rho e^{-i(\theta - \tau)} + \rho^2},$$

$$= \frac{1 - \rho^2}{1 - 2\rho \cos(\theta - \tau) + \rho^2}.$$

Hence

$$u(\rho, \theta) = \frac{1}{2\pi} \int_0^{2\pi} \frac{1 - \rho^2}{1 - 2\rho \cos(\theta - \tau) + \rho^2} f(\tau) d\tau, \tag{2.4.17}$$

which is called the **Poisson integral formula**. We will now show that it satisfies the boundary condition.

Now if $f(\theta) \equiv 1$ then $u(\rho, \theta) \equiv 1$ for $0 \leq \rho < 1$. Therefore

$$\frac{1}{2\pi} \int_0^{2\pi} \frac{1 - \rho^2}{1 - 2\rho \cos(\theta - \tau) + \rho^2} d\tau = 1.$$

Hence

$$f(\theta) = \frac{1}{2\pi} \int_0^{2\pi} \frac{1 - \rho^2}{1 - 2\rho \cos(\theta - \tau) + \rho^2} f(\theta) d\tau, \quad 0 \leq \rho < 1.$$

Therefore

$$u(\rho, \theta) - f(\theta) = \frac{1}{2\pi} \int_0^{2\pi} \frac{(1 - \rho^2)[f(\tau) - f(\theta)]}{1 - 2\rho \cos(\theta - \tau) + \rho^2} d\tau.$$

$f(\theta)$ is uniformly continuous on $[0, 2\pi]$. Hence for a given $\epsilon > 0$ there exists a $\delta(\epsilon) > 0$ such that whenever $|\theta - \tau| < \delta$, then

$$|f(\theta) - f(\tau)| < \epsilon.$$

Now if $|\theta - \tau| \geq \delta$, so that $\theta - \tau \neq 2n\pi, n = 0, 1, 2, \cdots$, then

$$\lim_{\rho \to 1^-} \frac{(1 - \rho^2)}{1 - 2\rho \cos(\theta - \tau) + \rho^2} = 0.$$

Therefore there exists a number $\rho_0$ such that if $\mid \theta - \tau \mid \geq \delta$, then

$$\frac{(1 - \rho^2)}{1 - 2\rho \cos(\theta - \tau) + \rho^2} < \epsilon \ \ \forall \ 0 \leq \rho_0 \leq \rho < 1.$$

Hence, for $\rho_0 \leq \rho < 1$, we have

$$\mid u(\rho, \theta) - f(\theta) \mid \ \leq \ \frac{1}{2\pi} \int_{\mid\theta-\tau\mid\geq\delta} \frac{(1 - \rho^2) \mid f(\theta) - f(\tau) \mid}{1 - 2\rho \cos(\theta - \tau) + \rho^2} d\tau$$

$$+ \frac{1}{2\pi} \int_{\mid\theta-\tau\mid<\delta} \frac{(1 - \rho^2) \mid f(\theta) - f(\tau) \mid}{1 - 2\rho \cos(\theta - \tau) + \rho^2} d\tau,$$

$$\leq \ \frac{1}{2\pi} 2\pi\epsilon\{2 \max_{0\leq\theta<2\pi} \mid f(\theta) \mid\} + \frac{\epsilon}{2\pi} 2\pi.$$

This implies that $\lim_{\rho\to 1-} u(\rho, \theta) = f(\theta)$ uniformly. Hence we have proved the following theorem.

**Theorem 2.4.6**: The solution for the Dirichlet problem for a circle of radius $a$ is given by the Poisson integral formula (2.4.17). $\qquad\square$

This solution is unique as already shown in Theorem 2.4.3. Hence the Dirichlet problem for a circle of radius $a$ is well posed as we have already proved that it depends continuously on the boundary data.

## 2.4.7   The Dirichlet Exterior Problem for a Circle

Here we want to find a harmonic function $u$ in the region $r > a$ if $u$ is given on $r = a$. Here $u$ must be bounded as $r \to \infty$. Such a general solution is

$$u(r, \theta) = \frac{a_0}{2} + \sum_{n=1}^{\infty} \left(\frac{r}{a}\right)^{-n} (a_n \cos n\theta + b_n \sin n\theta). \tag{2.4.18}$$

Applying the boundary condition $u(a, \theta) = f(\theta)$ , we obtain

$$f(\theta) = \frac{a_0}{2} + \sum_{n=1}^{\infty} (a_n \cos n\theta + b_n \sin n\theta).$$

Hence

$$a_n = \frac{1}{\pi} \int_0^{2\pi} f(\tau) \cos n\tau \ d\tau, \quad n = 0, 1, 2, \cdots,$$

$$b_n = \frac{1}{\pi} \int_0^{2\pi} f(\tau) \sin n\tau \ d\tau, \quad n = 1, 2, \cdots.$$

On substituting for $a_n$ and $b_n$ in Equation (2.4.18) and proceeding as in the previous section, we get

$$u(\rho,\theta) = \frac{1}{2\pi}\int_0^{2\pi}[1+2\sum_{n=1}^{\infty}\rho^{-n}\cos n(\theta-\tau)]f(\tau)d\tau,$$

$$= \frac{1}{2\pi}\int_0^{2\pi}\frac{\rho^2-1}{1-2\rho\cos(\theta-\tau)+\rho^2}f(\tau)d\tau, \qquad (2.4.19)$$

where $\rho = \dfrac{r}{a}$.

## 2.4.8   The Neumann Problem for a Circle

Solve
$$\nabla^2 u = 0, \quad r < a,$$

subject to the boundary condition

$$\frac{\partial u}{\partial n} = \frac{\partial u}{\partial r} = f(\theta) \quad \text{on} \quad r = a,$$

where
$$\int_0^{2\pi} f(\theta)d\theta = 0.$$

By the same argument as in Dirichlet's problem

$$u(r,\theta) = \frac{a_0}{2} + \sum_{n=1}^{\infty}\left(\frac{r}{a}\right)^n(a_n\cos n\theta + b_n\sin n\theta). \qquad (2.4.20)$$

Now applying the boundary condition, we get

$$\frac{\partial u}{\partial r}(a,\theta) = \sum_{n=1}^{\infty}\frac{n}{a}(a_n\cos n\theta + b_n\sin n\theta) = f(\theta).$$

So

$$a_n = \frac{a}{n\pi}\int_0^{2\pi}f(\tau)\cos n\tau d\tau, \quad n=1,2,3\cdots,$$

$$b_n = \frac{a}{n\pi}\int_0^{2\pi}f(\tau)\sin n\tau d\tau, \quad n=1,2,3\cdots.$$

$a_0$ is not determined.

Substituting for $a_n$ and $b_n$ in (2.4.20), we get

$$u(r,\theta) = \frac{a_0}{2} + \frac{a}{\pi} \int_0^{2\pi} \sum_{n=1}^{\infty} \frac{1}{n} \left(\frac{r}{a}\right)^n \cos n(\theta - \tau) f(\tau) d\tau.$$

The interchange of the integral with the summation is allowed here also for the same reasons as were explained in the Dirichlet problem. We shall use the following identity

$$-\frac{1}{2} \log[1 + \rho^2 - 2\rho \cos(\theta - \tau)] = \sum_{n=1}^{\infty} \frac{1}{n} \rho^n \cos n(\theta - \tau).$$

Then $u(r,\theta)$ becomes

$$u(r,\theta) = \frac{a_0}{2} - \frac{a}{2\pi} \int_0^{2\pi} \log[a^2 - 2ar \cos(\theta - \tau) + r^2] f(\tau) d\tau, \tag{2.4.21}$$

in which a constant factor $a^2$ in the argument of the logarithm was eliminated by virtue of the necessary condition for the Neumann problem.

As in the case of the Dirichlet problem, the exterior Neumann problem can be solved. In this case, the solution is given by

$$u(r,\theta) = \frac{a_0}{2} + \frac{a}{2\pi} \int_0^{2\pi} \log[a^2 - 2ar \cos(\theta - \tau) + r^2] f(\tau) d\tau.$$

## 2.4.9    The Dirichlet Problem for a Rectangle

Consider the following problem

$$\nabla^2 u = u_{xx} + u_{yy} = 0, \quad 0 < x < a, \ 0 < y < b, \tag{2.4.22}$$

with the boundary conditions

$$u(x,0) = f(x), \quad 0 \le x \le a, \tag{2.4.23}$$
$$u(x,b) = 0, \quad 0 \le x \le a, \tag{2.4.24}$$
$$u(0,y) = 0, \quad 0 \le y \le b, \tag{2.4.25}$$
$$u(a,y) = 0, \quad 0 \le y \le b. \tag{2.4.26}$$

**Note:** Here, we will derive the solution with one of the boundary conditions being non-homogeneous. By superposing the solutions thus obtained, i.e., by considering one of the boundary conditions to be non-homogeneous at a time, we can solve the general case as the equations are linear.

By assuming a variable separable solution of the form $u = X(x)Y(y)$, we get

$$X'' - \lambda X = 0,$$

$$Y'' + \lambda Y = 0.$$

The conditions (2.4.25) and (2.4.26) give $X(0) = X(a) = 0$.

Therefore the eigenvalues are $\lambda_n = -(n^2\pi^2)/a^2$ and the corresponding eigenfunctions are $X_n = B_n \sin\left(\dfrac{n\pi x}{a}\right)$. The corresponding values of $Y_n(y)$ are

$$Y_n(y) = E_n \sinh \frac{n\pi(y-b)}{a},$$

as we require $Y_n(b) = 0$ to satisfy the boundary condition (2.4.24). Therefore

$$u(x,y) = \sum_{n=1}^{\infty} X_n Y_n,$$

$$= \sum_{n=1}^{\infty} a_n \sin\left(\frac{n\pi x}{a}\right) \sinh \frac{n\pi(y-b)}{a},$$

where $a_n = B_n E_n$. Using the boundary condition (2.4.23) we get

$$f(x) = \sum_{n=1}^{\infty} a_n \sin\left(\frac{n\pi x}{a}\right) \sinh\left(\frac{-n\pi b}{a}\right).$$

Hence

$$a_n = -\frac{2}{a \sinh\left(\frac{n\pi b}{a}\right)} \int_0^a f(x) \sin\left(\frac{n\pi x}{a}\right) dx.$$

Therefore

$$u(x,y) = \sum_{n=1}^{\infty} a_n^* \frac{\sinh \frac{n\pi(y-b)}{a}}{\sinh \frac{n\pi b}{a}} \sin\left(\frac{n\pi x}{a}\right), \tag{2.4.27}$$

where

$$a_n^* = -\frac{2}{a} \int_0^a f(x) \sin\left(\frac{n\pi x}{a}\right) dx. \tag{2.4.28}$$

## 2.4.10   Harnack's Theorem

**Lemma 2.4.1**: Let $D$ be a bounded domain in $\mathbb{R}^2$, bounded by a smooth closed curve $B$. Let $\{u_n\}$ be a sequence of functions each of which is continuous on $\bar{D} = D \cup B$ and harmonic in $D$. If $\{u_n\}$ converges **uniformly** on $B$, then $\{u_n\}$ converges **uniformly** on $\bar{D}$.

**Proof**: By assumption, $\{u_n\}$ converges uniformly on $B$. Thus, given $\epsilon > 0$, $\exists\ N(\epsilon)$ such that

$$| u_m(x,y) - u_n(x,y) | < \epsilon \ \forall \ m,n > N(\epsilon), \ \forall\ (x,y) \in B.$$

But since $(u_m - u_n)$ is harmonic by the Maximum and Minimum principle,

$$| u_m(x,y) - u_n(x,y) | < \epsilon \ \forall \ m,n > N(\epsilon), \ \forall\ (x,y) \in \bar{D}.$$

Hence the result.                                                                    □

**Theorem 2.4.7: (Harnack's Theorem)** Let $D$ be a bounded domain in $\mathbb{R}^2$, bounded by a closed smooth curve $B$. Let $\{u_n\}$ be a sequence of functions, each of which is continuous on $\bar{D}$ and harmonic in $D$. If $\{u_n\}$ converges uniformly on $B$, then $\{u_n\}$ converges on $\bar{D}$ to a limit function that is continuous on $\bar{D}$ and harmonic in $D$.

**Proof**: By the previous lemma, $\{u_n\}$ converges uniformly on $\bar{D}$. Since on a closed bounded set, a uniformly convergent sequence of continuous functions converges to a function that is continuous on that set, $\{u_n\}$ converges to a function $u$ that is continuous on $\bar{D}$. In order to show that $u$ is harmonic in $D$, we show that it is given by the Poisson integral formula.

Let $(x, y) \in D$. Since $D$ is open, $\exists$ 'a' such that a circle with center $(x, y)$ and radius $a$ is contained in $D$. Any point on the circle will be $(x + a\cos\tau, y + a\sin\tau)$ for $0 \le \tau < 2\pi$.

Let $u_n(\tau) = u_n(x + a\cos\tau, y + a\sin\tau)$.

Since $u_n$ is harmonic inside the circle and continuous on the circle, we have

$$u_n(\xi, \eta) = \frac{1}{2\pi} \int_0^{2\pi} \frac{1 - \rho^2}{1 - 2\rho\cos(\theta - \tau) + \rho^2} u_n(\tau)d\tau,$$

where $\rho = (r/a)$ and $(\xi - x)^2 + (\eta - y)^2 = r^2 < a^2$. Hence we have

$$u(\xi, \eta) = \lim_{n \to \infty} u_n(\xi, \eta),$$
$$= \frac{1}{2\pi} \int_0^{2\pi} \frac{(1 - \rho^2)}{1 - 2\rho\cos(\theta - \tau) + \rho^2} \lim_{n \to \infty} u_n(\tau)d\tau,$$
$$= \frac{1}{2\pi} \int_0^{2\pi} \frac{(1 - \rho^2)}{1 - 2\rho\cos(\theta - \tau) + \rho^2} u(\tau)d\tau.$$

Observe that the limit and the integral have been interchanged in the previous equations due to the uniform convergence of the sequence $\{u_n\}$ to $u$. Hence $u$ is harmonic in the region $(x - \xi)^2 + (y - \eta)^2 < a^2$ for all $(\xi, \eta)$. Since $(x, y)$ is an arbitrary point of $D$, $u$ is harmonic in $D$. Hence the theorem. $\qquad\square$

## 2.4.11   Laplace's Equation — Green's Function

Let us recall the Green's identity given in (2.4.3)

$$\int_D (\psi \nabla^2 \phi - \phi \nabla^2 \psi) dS = \int_B (\psi \frac{\partial \phi}{\partial n} - \phi \frac{\partial \psi}{\partial n}) ds. \qquad (2.4.29)$$

Consider a point $P(x, y)$ inside a domain $S$ bounded by the smooth closed curve $B$ and let $\psi(x, y)$ be a harmonic function in $S$. Let $\Gamma$ be a circle inside $S$ with center $P$ and radius $\epsilon$. Now apply the identity (2.4.29) to the region $D$ bounded by the curves $B$ and $\Gamma$, where

$$\phi = \log \frac{1}{|\vec{r} - \vec{r'}|} ,$$

and

$$\vec{r} = x\vec{i} + y\vec{j}, \quad \vec{r'} = x'\vec{i} + y'\vec{j},$$

$(x', y')$ being any point in $D$. Observe that $\phi(x', y')$ is a harmonic function in $(x', y')$ in any domain that does not contain the point $(x, y)$. This function is called the fundamental solution of Laplace's equation in two dimensions.

Since both $\phi$ and $\psi$ are harmonic in $D$, observe that

$$\left( \int_\Gamma + \int_B \right) \left\{ \psi(x', y') \frac{\partial}{\partial n} \log \frac{1}{|\vec{r} - \vec{r'}|} - \log \frac{1}{|\vec{r} - \vec{r'}|} \frac{\partial \psi}{\partial n} \right\} ds' = 0,$$

where the integrand is the same for both the integrals. Observe that

$$\int_\Gamma \psi \frac{\partial}{\partial n} \log \frac{1}{|\vec{r} - \vec{r'}|} ds' = 2\pi \psi(x, y) + \mathrm{O}(\epsilon),$$

$$\left| \int_\Gamma \log \frac{1}{|\vec{r} - \vec{r'}|} \frac{\partial \psi}{\partial n} ds' \right| \leq 2\pi M \epsilon \, |\log \epsilon \,|,$$

where $M$ is the upper bound of $|\dfrac{\partial \psi}{\partial n}|$ on $\Gamma$. Therefore as $\epsilon \to 0$

$$\psi(x, y) = \frac{1}{2\pi} \int_B \left\{ \log \frac{1}{|\vec{r} - \vec{r'}|} \frac{\partial}{\partial n} \psi(x', y') - \psi(x', y') \frac{\partial}{\partial n} \log \frac{1}{|\vec{r} - \vec{r'}|} \right\} ds'. \qquad (2.4.30)$$

Let us define a function

$$G(x, y; x', y') = W(x, y; x', y') + \log \frac{1}{|\vec{r} - \vec{r'}|} ,$$

where $W$ is a harmonic function of $(x', y')$ in $S$ and $W(x, y; x', y') = \log |\vec{r} - \vec{r'}|$ on $B$. Such a function $G$, if it exists, is called the Green's function for Laplace's equation. Further, if $\psi = f(x, y)$ on $B$, we get

$$\psi(x, y) = -\frac{1}{2\pi} \int_B \psi(x', y') \frac{\partial}{\partial n} G(x, y; x', y') ds',$$

$$= -\frac{1}{2\pi} \int_B f(x', y') \frac{\partial}{\partial n} G(x, y; x', y') ds'. \qquad (2.4.31)$$

**Note**: (2.4.31) results as follows. Applying (2.4.29) for $\psi$ and $W$ we get

$$\frac{1}{2\pi} \int_B (\psi \frac{\partial W}{\partial n} - W \frac{\partial \psi}{\partial n}) ds' = 0. \qquad (2.4.32)$$

Subtracting (2.4.32) from (2.4.30) we get

$$\psi(x, y) = \frac{1}{2\pi} \int_B \left[ \frac{\partial \psi}{\partial n} \left( W + \log \frac{1}{|\vec{r} - \vec{r'}|} \right) \right.$$

$$\left. -\psi \frac{\partial}{\partial n} \left( W + \log \frac{1}{|\vec{r} - \vec{r'}|} \right) \right] ds'. \qquad (2.4.33)$$

Since $G = (W + \log \frac{1}{|\vec{r} - \vec{r'}|}) = 0$ on $B$, Equation (2.4.33) reduces to (2.4.31).

We shall use the formula (2.4.31) to find the solution of the Dirichlet problem for the upper half plane and a circle by choosing a suitable $G$ in each case, in the next two sections.

## 2.4.12    The Dirichlet Problem for a Half Plane

As was already discussed in Section 2.4.4, the problem involves solving

$$\nabla^2 \psi = 0, \quad y > 0, \quad -\infty < x < \infty, \qquad (2.4.34)$$

with the conditions

$$\psi(x, 0) = f(x), \quad -\infty < x < \infty, \qquad (2.4.35)$$

$$\psi(x, y) \to 0 \quad \text{as} \quad y \to \infty.$$

Let $P(x, y)$ be any point in the domain $y > 0$. Let $Q$ be the point with coordinates $(x', y')$ and $P'$ be $(x, -y)$. The suitable $G$ here is

$$G(x, y; x', y') = \log \frac{QP'}{QP},$$

$$= \frac{1}{2} \log \frac{(x - x')^2 + (y + y')^2}{(x - x')^2 + (y - y')^2}. \tag{2.4.36}$$

Observe that the $G$ given above is a harmonic function of $(x', y')$ in the region $y' > 0$ except at $P(x, y)$ and it vanishes on the boundary $y' = 0$. Also

$$\frac{\partial G}{\partial n} = -\left. \frac{\partial G}{\partial y'} \right|_{y'=0} = \frac{-2y}{(x - x')^2 + y^2}.$$

Therefore, from Equation (2.4.31), we have

$$\psi(x, y) = \frac{y}{\pi} \int_{-\infty}^{\infty} \frac{f(x') dx'}{(x - x')^2 + y^2}. \tag{2.4.37}$$

This is in agreement with the formula given in Equation (2.4.10).

## 2.4.13   The Dirichlet Problem for a Circle

As was already discussed in Section 2.4.6, the problem involves solving

$$\nabla^2 \psi = 0, \qquad r < a, \tag{2.4.38}$$

with the condition that

$$\psi(a, \theta) = f(\theta). \tag{2.4.39}$$

Let $P$ be the point $(r, \theta)$ and $Q$ be the point $(r', \theta')$.
Let $P'$ be the inverse point of $P$ with respect to the circle $r = a$. Therefore $P'$ is given by $(a^2/r, \theta)$. Then the suitable $G$ is

$$G(r, \theta; r', \theta') = \log \left( \frac{r \cdot P'Q}{a \cdot PQ} \right).$$

Observe that such a $G$ is harmonic inside the circle except at the point $P$. In addition, $G$ vanishes on the circle $r' = a$. So written in terms of $(r, \theta)$ and $(r', \theta')$, $G$ is of the form

$$G(r, \theta; r', \theta') = \frac{1}{2} \log \frac{a^2 + r^2 r'^2/a^2 - 2rr' \cos(\theta' - \theta)}{r'^2 + r^2 - 2rr' \cos(\theta' - \theta)}. \tag{2.4.40}$$

On the boundary $r' = a$,

$$\frac{\partial G}{\partial n} = \frac{\partial G}{\partial r'}\bigg|_{r'=a} = \frac{-(a^2 - r^2)}{a[a^2 - 2ar\cos(\theta' - \theta) + r^2]} \ ,$$

so that

$$\psi(r,\theta) = \frac{(a^2 - r^2)}{2\pi} \int_0^{2\pi} \frac{f(\theta')d\theta'}{a^2 - 2ar\cos(\theta' - \theta) + r^2} \ . \tag{2.4.41}$$

This formula is the **Poisson integral formula**, which we derived earlier by summing the series (refer to Equation (2.4.17)).

**Note 1**: $G(P; Q) = G(Q; P)$ in the previous two cases. In fact, this result is true in general. That is, Green's function is symmetric.

**Note 2**: If $\psi(x, y)$ satisfies

$$\nabla^2\psi = -F(x, y),$$

then instead of (2.4.31), we get

$$\psi(x, y) = \int\int_S G(x, y; x', y')F(x', y')dS'$$
$$- \frac{1}{2\pi} \int_B f(x', y')\frac{\partial}{\partial n}G(x, y; x', y')ds'.$$

## 2.5   Heat Conduction Problem

### 2.5.1   Heat Conduction — Infinite Rod Case

Consider the following heat conduction problem in an infinite rod with the following assumptions:

- The position of the rod coincides with the $x$-axis, and the rod is homogeneous.

- The rod is sufficiently thin so that the heat is uniformly distributed over its cross section at a given time $t$.

- The surface of the rod is insulated to prevent any loss of heat through the boundary.

- $u(x, t)$ is the temperature at the point $x$ at time $t$.

Then the problem is to solve

$$u_t = ku_{xx}, \qquad -\infty < x < \infty, \quad t > 0, \tag{2.5.1}$$

$$u(x, 0) = f(x), \qquad -\infty < x < \infty. \tag{2.5.2}$$

Suppose the Fourier transform of $u(x, t)$ is $U(\alpha, t)$, i.e.,

$$\mathcal{F}[u(x, t)] = U(\alpha, t) = \frac{1}{\sqrt{2\pi}} \int_{-\infty}^{\infty} u(x, t) e^{i\alpha x} dx.$$

Taking the Fourier transform of (2.5.1) (refer to Appendix A), and assuming that $u$, $u_x \to 0$ as $|x| \to \infty$, we get

$$U_t + k\alpha^2 U = 0.$$

Its solution is given by

$$U(\alpha, t) = A(\alpha) e^{-\alpha^2 kt},$$

where $A(\alpha)$ is an arbitrary function to be determined from the initial condition as follows:

$$U(\alpha, 0) = \mathcal{F}[u(x, 0)] \quad = \quad \frac{1}{\sqrt{2\pi}} \int_{-\infty}^{\infty} u(x, 0) e^{i\alpha x} dx,$$

$$= \quad \frac{1}{\sqrt{2\pi}} \int_{-\infty}^{\infty} f(x) e^{i\alpha x} dx = F(\alpha), \text{ (say)}.$$

Hence

$$U(\alpha, t) = F(\alpha) e^{-\alpha^2 kt}.$$

Hence by the Convolution theorem (refer to Theorem A2 in Appendix A)

$$u(x, t) = f(x) * \mathcal{F}^{-1}(e^{-\alpha^2 kt}),$$

$$= \frac{1}{2\sqrt{\pi kt}} \int_{-\infty}^{\infty} f(\xi) \exp\left(-\frac{(x - \xi)^2}{4kt}\right) d\xi. \tag{2.5.3}$$

Equation (2.5.3) gives the required solution of (2.5.1) and (2.5.2). Now, consider the case $k = 1$ and

$$f(x) = \begin{cases} 0, & x < 0, \\ a, & x > 0. \end{cases} \tag{2.5.4}$$

$$u(x, t) = \frac{a}{2\sqrt{\pi t}} \int_{0}^{\infty} \exp\left(-\frac{(x - \xi)^2}{4t}\right) d\xi. \tag{2.5.5}$$

Put $\eta = \dfrac{(\xi - x)}{2\sqrt{t}}$.

$$u(x, t) = \frac{a}{\sqrt{\pi}} \int_{-x/2\sqrt{t}}^{\infty} e^{-\eta^2} d\eta \, ,$$

$$= \frac{a}{\sqrt{\pi}} \left[ \int_{-x/2\sqrt{t}}^{0} e^{-\eta^2} d\eta + \int_{0}^{\infty} e^{-\eta^2} d\eta \right] \, ,$$

$$= \frac{a}{2} + \frac{a}{\sqrt{\pi}} \int_{0}^{x/2\sqrt{t}} e^{-\eta^2} d\eta \, ,$$

$$= \frac{a}{2} \left[ 1 + \text{erf} \left( \frac{x}{2\sqrt{t}} \right) \right] \, ,$$

where $\text{erf}(x)$ is the error function.

## 2.5.2   Heat Conduction — Finite Rod Case

Let us now consider the problem of heat conduction in a finite rod of length $l$ with the same assumptions regarding the nature of the rod as in the infinite rod case. The governing equation along with the initial and the boundary conditions takes the following form

$$u_t = k u_{xx}, \qquad 0 < x < l, \quad t > 0, \tag{2.5.6}$$
$$u(0, t) = u(l, t) = 0, \quad t > 0, \tag{2.5.7}$$
$$u(x, 0) = f(x), \qquad 0 \le x \le l. \tag{2.5.8}$$

Let us assume the solution of Equation (2.5.6) as

$$u(x, t) = X(x)T(t).$$

Then, Equation (2.5.6) gives

$$\frac{X''}{X} = \frac{T'}{kT} = \text{constant} = \lambda, \text{ (say)}.$$

Once again we observe that $\lambda$ must be negative and therefore equal to $-\alpha^2$. Hence

$$X'' + \alpha^2 X = 0,$$

$$T' + \alpha^2 kT = 0.$$

Therefore

$$X(x) = A\cos\alpha x + B\sin\alpha x.$$

Equation (2.5.7) implies that $X(0) = X(l) = 0$, and

$$X(0) = 0 \Rightarrow A = 0,$$
$$X(l) = 0 \Rightarrow B\sin\alpha l = 0.$$

Since $B = 0$ yields only trivial solutions, it follows that $\sin\alpha l = 0$. Therefore $\alpha_n = (n\pi)/l$, $n = 1, 2, 3, \cdots$. Hence

$$X_n(x) = B_n \sin\left(\frac{n\pi x}{l}\right),$$

and

$$T_n(t) = C_n \exp(-\frac{n^2\pi^2 kt}{l^2}),$$

are the corresponding eigenfunctions. Therefore

$$u_n(x, t) = a_n \exp(-\frac{n^2\pi^2 kt}{l^2})\sin\left(\frac{n\pi x}{l}\right).$$

Therefore by the principle of superposition

$$u(x, t) = \sum_{n=1}^{\infty} a_n \exp(-\frac{n^2\pi^2 kt}{l^2})\sin\left(\frac{n\pi x}{l}\right). \qquad (2.5.9)$$

Now $u(x, 0) = f(x)$ implies

$$f(x) = \sum_{n=1}^{\infty} a_n \sin\left(\frac{n\pi x}{l}\right), \quad 0 \le x \le l, \qquad (2.5.10)$$

and

$$a_n = \frac{2}{l}\int_0^l f(x)\sin\left(\frac{n\pi x}{l}\right) dx. \qquad (2.5.11)$$

The function $u(x, t)$ given in (2.5.9) represents the solution of Equation (2.5.6) if $f(x)$ is twice continuously differentiable and $f'''$ is piecewise continuous. In fact, this ensures, as stated earlier in Section 2.3.3, that the series given in (2.5.9) can be differentiated term-by-term twice with respect to $x$.

Observe that the series given in Equation (2.5.9) converges uniformly if the series given in Equation (2.5.10) converges uniformly.

**Exercise 2.5.1**: Solve

$$u_t = u_{xx}, \quad 0 < x < l, \quad t > 0,$$
$$u(0,t) = u(l,t) = 0,$$
$$u(x,0) = x(l - x), \quad 0 \le x \le l.$$

**Ans.**: The solution of this problem is given by (2.5.9), where $a_n = (8l^2)/n^3\pi^3$ if $n$ is odd and $a_n = 0$ if $n$ is even.

**Theorem 2.5.1: Uniqueness of the solution**

The solution $u(x,t)$ of the differential equation

$$u_t - ku_{xx} = F(x,t), \quad 0 < x < l, \quad t > 0, \tag{2.5.12}$$

satisfying the initial condition

$$u(x,0) = f(x), \quad 0 \le x \le l, \tag{2.5.13}$$

and the boundary conditions

$$u(0,t) = u(l,t) = 0, \quad t \ge 0, \tag{2.5.14}$$

is unique.

**Proof**: Let $u_1(x,t)$ and $u_2(x,t)$ be two solutions of this problem. Then $v = (u_1 - u_2)$ satisfies

$$v_t = kv_{xx}, \quad 0 < x < l, \quad t > 0, \tag{2.5.15}$$
$$v(0,t) = v(l,t) = 0, \quad t \ge 0, \tag{2.5.16}$$
$$v(x,0) = 0, \quad 0 \le x \le l. \tag{2.5.17}$$

Let us define a function

$$E(t) = \frac{1}{2k} \int_0^l v^2(x,t)dx.$$

Therefore $E \ge 0$. On differentiating this function with respect to $t$, we get

$$\frac{dE}{dt} = \frac{1}{k} \int_0^l vv_t dx,$$
$$= \int_0^l vv_{xx}dx \text{ (due to Equation (2.5.15))},$$
$$= vv_x \big|_0^l - \int_0^l v_x^2 dx \text{ (on integrating by parts)}.$$

Since $v(0,t) = v(l,t) = 0$, we find that

$$\frac{dE}{dt}(t) = -\int_0^l v_x^2 dx \leq 0.$$

Therefore $E$ is a decreasing function of $t$.

From the condition $v(x,0) = 0$, we have $E(0) = 0$.

Therefore $E(t) \leq 0$ for all $t > 0$. But $E(t)$, by definition, is non-negative. Therefore

$$E(t) \equiv 0 \quad \forall \quad t > 0 \Rightarrow v(x,t) \equiv 0 \quad \text{in} \ \ 0 \leq x \leq l, \ \ t \geq 0.$$

Therefore $u_1 \equiv u_2$. Hence the solution is unique. $\qquad\qquad\square$

**Note**: The solution of the problem of heat conduction in a finite rod stated in Equations (2.5.6)–(2.5.8), being a special case of Theorem 2.5.1, is also unique.

Suppose the temperature at one end of the rod is maintained to be constant, i.e.,

$$u(l,t) = u_0 \quad \forall \quad t > 0.$$

Put $v = u - (u_0 x)/l$. Then $v$ satisfies

$$v_t = kv_{xx}, \quad 0 < x < l, \ \ t > 0, \tag{2.5.18}$$

$$v(0,t) = v(l,t) = 0, \ \ t > 0, \tag{2.5.19}$$

$$v(x,0) = f(x) - \frac{u_0 x}{l}, \quad 0 \leq x \leq l. \tag{2.5.20}$$

Then the solution can now be found as in (2.5.9), (2.5.10), and (2.5.11).

**Exercise 2.5.2**: Let $v(\eta) = c\int_0^\eta \frac{e^{-s/4}}{\sqrt{s}} ds + d$, where $c$ and $d$ are constants. Define $u(x,t) = v(\frac{x^2}{t})$. Show that $u(x,t)$ satisfies the heat conduction equation (2.5.1).

## 2.6  Duhamel's Principle

In Sections 2.3 and 2.5, we solved the wave and heat equations. The partial differential equations considered in both of these cases are homogeneous. Duhamel suggested a principle, which is known after his name, to construct solutions of non-homogeneous differential equations. This technique can be applied to both hyperbolic and parabolic equations for the Cauchy problems and initial-boundary value problems. We will demonstrate this technique for the wave and heat conduction equations.

## 2.6.1   Wave Equation

Let us consider the non-homogeneous wave equation

$$u_{tt} - c^2 u_{xx} = F(x,t), \quad -\infty < x < \infty, \ t > 0, \tag{2.6.1}$$

with the homogeneous initial conditions

$$u(x,0) = u_t(x,0) = 0, \quad -\infty < x < \infty. \tag{2.6.2}$$

We consider the function $v(x,t;\tau)$, which satisfies the following equation with respect to $x$ and $t$ for $t > \tau$,

$$v_{tt} - c^2 v_{xx} = 0, \quad -\infty < x < \infty, \ t > \tau > 0, \tag{2.6.3}$$

and the following conditions at $t = \tau$

$$v(x,\tau;\tau) = 0, \quad v_t(x,\tau;\tau) = F(x,\tau). \tag{2.6.4}$$

The solution for this problem is (refer to d'Alembert's solution)

$$v(x,t;\tau) = \frac{1}{2c} \int_{x-c(t-\tau)}^{x+c(t-\tau)} F(s,\tau)ds. \tag{2.6.5}$$

Consider

$$u(x,t) = \int_0^t v(x,t;\tau)d\tau. \tag{2.6.6}$$

We will now show that $u$ is the solution we are looking for.

$$u_t = v(x,t;t) + \int_0^t v_t(x,t;\tau)d\tau,$$

$$= \int_0^t v_t(x,t;\tau)d\tau \quad (\text{since } v(x,t;t) = 0),$$

$$u_{tt} = v_t(x,t;t) + \int_0^t v_{tt}(x,t;\tau)d\tau,$$

$$= F(x,t) + \int_0^t v_{tt}(x,t;\tau)d\tau,$$

$$u_{xx} = \int_0^t v_{xx}(x,t;\tau)d\tau.$$

Therefore

$$u_{tt} - c^2 u_{xx} = F(x,t) + \int_0^t (v_{tt} - c^2 v_{xx})d\tau,$$
$$= F(x,t).$$

Also observe that $u(x,t)$ satisfies the conditions (2.6.2).
If the initial data (2.6.2) is replaced by

$$u(x,0) = f(x), \quad u_t(x,0) = g(x), \quad -\infty < x < \infty,$$

then the solution of (2.6.1) is found by superposing $u$ from (2.6.6) on d'Alembert's solution, i.e.,

$$u(x,t) = \frac{1}{2}[f(x+ct) + f(x-ct)] + \frac{1}{2c}\int_{x-ct}^{x+ct} g(s)ds$$
$$+\frac{1}{2c}\int_0^t \int_{x-c(t-\tau)}^{x+c(t-\tau)} F(s,\tau)ds d\tau. \tag{2.6.7}$$

**Note**: Observe that in the non-homogeneous case, the domain of dependence is the entire triangular region given in Figure 2.3.2 bounded by the two characteristics and the $x$-axis unlike in the homogeneous case where it is only the segment $AB$ on the $x$-axis.

**Exercise 2.6.1**: Solve

$$u_{tt} - c^2 u_{xx} = F(x,t), \quad 0 < x < l, \ t > 0,$$

$$u(x,0) = f(x), \quad 0 < x < l,$$
$$u_t(x,0) = g(x), \quad 0 < x < l,$$
$$u(0,t) = u(l,t) = 0, \quad t > 0,$$

by making use of Duhamel's principle.
**Solution**:

$$u(x,t) = \sum_{n=1}^{\infty} \left( a_n \cos(\frac{n\pi ct}{l}) + b_n \sin(\frac{n\pi ct}{l}) \right.$$
$$\left. + \int_0^t c_n(\tau) \sin(\frac{n\pi c(t-\tau)}{l})d\tau \right) \sin(\frac{n\pi x}{l}),$$

where

$$a_n = \frac{2}{l}\int_0^l f(x)\sin(\frac{n\pi x}{l})dx, \quad b_n = \frac{2}{n\pi c}\int_0^l g(x)\sin(\frac{n\pi x}{l})dx,$$

$$c_n(\tau) = \frac{2}{n\pi c}\int_0^l F(x,\tau)\sin(\frac{n\pi x}{l})dx.$$

**Note**: We have already shown that the solution of the previous problem, if it exists, is unique in Theorem 2.3.2.

## 2.6.2   Heat Conduction Equation

Let us now consider the heat conduction equation in an infinite rod with a heat source. The governing equation is

$$u_t - ku_{xx} = F(x,t), \quad -\infty < x < \infty, \quad t > 0, \tag{2.6.8}$$

with the initial condition

$$u(x,0) = 0, \quad -\infty < x < \infty. \tag{2.6.9}$$

Let us consider the function $v(x,t;\tau)$ satisfying

$$v_t - kv_{xx} = 0, \quad -\infty < x < \infty, \quad t > \tau > 0, \tag{2.6.10}$$

and the following condition at $t = \tau$

$$v(x,\tau;\tau) = F(x,\tau). \tag{2.6.11}$$

Then $v$ is given by

$$v(x,t;\tau) = \frac{1}{2\sqrt{\pi k(t-\tau)}}\int_{-\infty}^{\infty} F(\xi,\tau)\exp(-\frac{(x-\xi)^2}{4k(t-\tau)})d\xi. \tag{2.6.12}$$

Consider

$$u(x,t) = \int_0^t v(x,t;\tau)d\tau. \tag{2.6.13}$$

Then

$$u_t(x,t) = v(x,t;t) + \int_0^t v_t(x,t;\tau)d\tau,$$

$$= F(x,t) + \int_0^t v_t(x,t;\tau)d\tau.$$

Therefore

$$u_t - ku_{xx} = F(x,t) + \int_0^t [v_t(x,t;\tau) - kv_{xx}(x,t;\tau)]d\tau,$$

$$= F(x,t),$$

and $\qquad u(x,0) = 0.$

Hence $u$ given in (2.6.13) is the required solution of (2.6.8) and (2.6.9). Now suppose we consider the initial condition

$$u(x,0) = f(x),$$

instead of (2.6.9). Then the solution of (2.6.8) is

$$u(x,t) = \int_0^t \frac{1}{2\sqrt{\pi k(t-\tau)}} \int_{-\infty}^\infty F(\xi,\tau) \exp(-\frac{(x-\xi)^2}{4k(t-\tau)})d\xi d\tau$$

$$+ \frac{1}{2\sqrt{\pi kt}} \int_{-\infty}^\infty f(\xi) \exp(-\frac{(x-\xi)^2}{4kt})d\xi. \qquad (2.6.14)$$

**Exercise 2.6.2**: Solve

$$u_t - ku_{xx} = F(x,t), \quad 0 < x < l, \ t > 0,$$

$$u(x,0) = f(x), \text{ for } 0 < x < l, \text{ and } u(0,t) = u(l,t) = 0, \text{ for } t > 0.$$

**Note**: We have shown in Theorem 2.5.1 that the solution of the previous problem, if it exists, is unique.

**Solution** : $\ u(x,t) \ = \ \displaystyle\sum_{n=1}^\infty \left[ a_n \exp(-\frac{n^2\pi^2 kt}{l^2}) \right.$

$$\left. + \int_0^t c_n(\tau) \exp(-\frac{n^2\pi^2 k(t-\tau)}{l^2})d\tau \right] \sin(\frac{n\pi x}{l}),$$

where

$$a_n \ = \ \frac{2}{l} \int_0^l f(x) \sin(\frac{n\pi x}{l})dx, c_n(\tau) = \frac{2}{l} \int_0^l F(x,\tau) \sin(\frac{n\pi x}{l})dx.$$

## 2.7    Classification in the Case of $n$-Variables

In Section 2.2, we have considered the classification of a second order semi-linear p.d.e. in the case of two independent variables. We now discuss the classification of a second order semi-linear p.d.e. in the case of $n$ independent variables. Consider the semi-linear p.d.e.

$$L[u] = \sum_{i=1}^{n} \sum_{j=1}^{n} a^{ij} \frac{\partial^2 u}{\partial x^i \partial x^j} = f(x^i, u, \frac{\partial u}{\partial x^i}), \qquad (2.7.1)$$

where $x^i$ are independent variables and $u$ is the dependent variable. The coefficients $a^{ij}$ of the second order derivatives determine the nature of the differential equation. Hence the nature of the partial differential equation is decided by the nature of the operator $L$.

Let us transform the variables from $x^i$ to $x'^i$. Then we have

$$\frac{\partial u}{\partial x^i} = \sum_{l=1}^{n} \frac{\partial u}{\partial x'^l} \frac{\partial x'^l}{\partial x^i},$$

$$\frac{\partial^2 u}{\partial x^i \partial x^j} = \sum_{l=1}^{n} \sum_{m=1}^{n} \frac{\partial^2 u}{\partial x'^l \partial x'^m} \frac{\partial x'^l}{\partial x^i} \frac{\partial x'^m}{\partial x^j} + \sum_{l=1}^{n} \frac{\partial^2 x'^l}{\partial x^i \partial x^j} \frac{\partial u}{\partial x'^l}.$$

Hence Equation (2.7.1) transforms into

$$L'[u] = \sum_{l=1}^{n} \sum_{m=1}^{n} a'^{lm} \frac{\partial^2 u}{\partial x'^l \partial x'^m} = g(x'^k, u, \frac{\partial u}{\partial x'^k}), \qquad (2.7.2)$$

where

$$a'^{lm} = \sum_{i=1}^{n} \sum_{j=1}^{n} a^{ij} \frac{\partial x'^l}{\partial x^i} \frac{\partial x'^m}{\partial x^j}. \qquad (2.7.3)$$

From Equation (2.7.3), we can observe the tensor character of $a^{ij}$. They transform like the components of a contravariant tensor of rank two.

Without any loss of generality, we can assume that $a^{ij} = a^{ji}$, since the mixed derivatives are symmetric as $u \in C^2$. Observe that

$$\mid a'^{lm} \mid = \mid a^{ij} \frac{\partial x'^l}{\partial x^i} \frac{\partial x'^m}{\partial x^j} \mid,$$

$$= \mid a^{ij} \frac{\partial x'^l}{\partial x^i} \mid\mid \frac{\partial x'^m}{\partial x^j} \mid,$$

$$= \mid a^{ij} \mid\mid \frac{\partial x'^l}{\partial x^i} \mid\mid \frac{\partial x'^m}{\partial x^j} \mid,$$

$$= \mid a^{ij} \mid\mid \frac{\partial x'}{\partial x} \mid^2, \tag{2.7.4}$$

where

$$\mid \frac{\partial x'}{\partial x} \mid = \frac{\partial(x'^1, x'^2, \cdots, x'^m)}{\partial(x^1, x^2, \cdots, x^m)}. \tag{2.7.5}$$

If the transformation is admissible, then $\mid \frac{\partial x'}{\partial x} \mid \neq 0$.

Let $P$ be any point in the domain. Let $\xi$ be a covariant vector representing a surface element at the point $P$.

Then the quadratic form $Q(\xi) = \sum_{i=1}^{n} \sum_{j=1}^{n} a^{ij} \xi_i \xi_j$, which is evidently an invariant, is known as the characteristic form relative to the operator $L$.

Let $\phi(x^i) = 0$ be a surface passing through $P$. Let us choose a new coordinate system of which one is $x'^1 = \phi(x^i)$. Then the coefficient of $\frac{\partial^2 u}{\partial x'^1 \partial x'^1}$ in $L'[u]$ is

$$a'^{11} = \sum_{i=1}^{n} \sum_{j=1}^{n} a^{ij} \frac{\partial \phi}{\partial x^i} \frac{\partial \phi}{\partial x^j} = Q(\phi). \tag{2.7.6}$$

Should this coefficient be zero, the surface element $\xi_i = \frac{\partial \phi}{\partial x^i}$ is said to be a characteristic and $\phi(x^i) = 0$ is said to be a characteristic surface and $L[u]$ can be calculated at $P$ if the values of $u$ and of the first derivatives of $u$ on the surface $\phi(x^i) = 0$ are known.

In a sense, therefore, $L$ degenerates to an operator of the first order on a characteristic surface.

When $L$ has no characteristic surface element at $P$, it is said to be elliptic. Clearly $L$ is elliptic at $P$ if and only if the quadratic form $Q(\phi)$ is definite. (We can assume it to be positive definite.)

From matrix theory, we know that the quadratic form is positive definite if and only if all the eigenvalues are positive and in which case the matrix is non-singular.

The matrix of the coefficient $a^{ij}$ may, however, be non-singular without the characteristic form being definite, i.e., in the case when the eigenvalues are all different from zero, but they may be of either sign (positive or negative). In this case, we say that the operator $L$ is hyperbolic (i.e., non-singular, indefinite characteristic form). Finally, if the matrix of coefficients is singular so that the determinant $\mid a^{ij} \mid$ is zero (i.e., at least one eigenvalue is zero), then $L$ is said to be parabolic. An operator is said to be parabolic if the differential operator of the second order in it can be expressed in terms of fewer than $n$ variables.

From (2.7.3) we can conclude that if $\mid a^{ij} \mid$ is different from zero in one coordinate system, it is so in every other admissible coordinate system. Conversely, the vanishing of $\mid a^{ij} \mid$ in one system must imply its vanishing in every other admissible system. That is, an equation that is parabolic in one coordinate system is parabolic in every other admissible system.

Again, regarding the distinction between elliptic and hyperbolic equations, we see that the definite or indefinite character of the invariant quadratic form $Q(\phi)$ is certainly independent of a particular coordinate system.

This establishes the invariant nature of the classification.

This can also be seen from the relation between the matrix $(a^{ij})$ and $(a'^{ij})$.

$$(a'^{ij}) = J(a^{ij})J^{\star},$$

where $J = (\dfrac{\partial x'}{\partial x})$ and $J^{\star}$ is the transpose of $J$.

It is well known that for such matrices the number of positive, negative, and zero eigenvalues is the same.

Let us now consider the case in which $n = 2$ (also refer to Section 2.2).

$$L[u] = Ru_{xx} + Su_{xy} + Tu_{yy},$$
$$= Ru_{xx} + \frac{1}{2}Su_{xy} + \frac{1}{2}Su_{yx} + Tu_{yy}. \tag{2.7.7}$$

Then

$$(a^{ij}) = \begin{pmatrix} R & S/2 \\ S/2 & T \end{pmatrix},$$

$$\mid a^{ij} \mid = \frac{1}{4}(4RT - S^2). \tag{2.7.8}$$

Therefore if $S^2 - 4RT = 0$, the operator is of parabolic type.
Let us consider the eigenvalues of $(a^{ij})$.

$$\begin{vmatrix} R - \lambda & S/2 \\ S/2 & T - \lambda \end{vmatrix} = 0,$$

$$4\lambda^2 - 4(R + T)\lambda + 4RT - S^2 = 0.$$

The previous quadratic equation in $\lambda$ will have both the roots (eigenvalues) of the same sign if $4RT - S^2 > 0$ and opposite sign if $4RT - S^2 < 0$. Hence the operator is of elliptic type if $S^2 - 4RT < 0$ and is of hyperbolic type if $S^2 - 4RT > 0$. Further,

$$a'^{11} = A(\xi_x, \xi_y), \quad a'^{22} = A(\eta_x, \eta_y),$$

and

$$a'^{21} = a'^{12} = B(\xi_x, \xi_y; \eta_x, \eta_y).$$

Therefore

$$\mid a'^{ij} \mid = A(\xi_x, \xi_y)A(\eta_x, \eta_y) - B^2.$$

We know

$$\mid a^{ij} \mid = \begin{vmatrix} R & S/2 \\ S/2 & T \end{vmatrix} = \frac{1}{4}(4RT - S^2).$$

From Equation (2.7.4), we get

$$A(\xi_x, \xi_y)A(\eta_x, \eta_y) - B^2 = \frac{1}{4}(4RT - S^2)(\xi_x\eta_y - \xi_y\eta_x)^2, \tag{2.7.9}$$

for

$$J = \begin{vmatrix} \xi_x & \xi_y \\ \eta_x & \eta_y \end{vmatrix}. \tag{2.7.10}$$

Note that Equation (2.7.9) is the same as Equation (2.2.4).
**Example 2.7.1:**

1. $u_{xx} + u_{yy} + u_{zz} = 0$ is of elliptic type. This is called the Laplace's equation in three dimensions and its solution is called a potential function or harmonic function ($n = 3$).

2. $u_t - k(u_{xx} + u_{yy} + u_{zz}) = 0$ is of parabolic type. This is called the heat equation in three dimensions ($n = 4$).

3. $u_{tt} - c^2(u_{xx} + u_{yy} + u_{zz}) = 0$ is of hyperbolic type. This is called the wave equation in three dimensions $(n = 4)$.    □

## Exercise 2.7.1:

1. Classify the following equations into hyperbolic, parabolic, or elliptic type.

   (a) $u_{xx} + 2u_{xy} + u_{yy} + 2u_{zz} - (1 + xy)u = 0$.
   **Ans.:** Parabolic.

   (b) $u_{xx} + 2u_{yz} + \cos x u_z - e^{y^2} u = \cosh z$.
   **Ans.:** Hyperbolic.

   (c) $7u_{xx} - 10u_{xy} - 22u_{yz} + 7u_{yy} - 16u_{xz} - 5u_{zz} = 0$.
   **Ans.:** Hyperbolic.

   (d) $e^z u_{xy} - u_{xx} = \log(x^2 + y^2 + z^2 + 1)$.
   **Ans.:** Parabolic.

   (e) $u_{xx} + 2(1 + \alpha y)u_{yz} = 0$.
   **Ans.:** If $\alpha \neq 0$, then the equation is hyperbolic provided $y \neq -\dfrac{1}{\alpha}$ and parabolic on the plane $y = -\dfrac{1}{\alpha}$. If $\alpha = 0$, it is hyperbolic everywhere.

2. Determine the regions where the following equation is of hyperbolic, elliptic, or parabolic type.
$$u_{xx} - 2x^2 u_{xz} + u_{yy} + u_{zz} = 0.$$
**Ans.:** Hyperbolic if $| x | > 1$, parabolic if $| x | = 1$, and elliptic if $| x | < 1$.

3. Show that $u(x, y, t)$ given by
$$u = \sum_{n=1}^{\infty} \sum_{m=1}^{\infty} (a_{nm} \cos(c\lambda_{nm}t) + b_{nm} \sin(c\lambda_{nm}t)) \sin\left(\frac{n\pi x}{a}\right) \sin\left(\frac{m\pi y}{b}\right),$$

satisfies the two-dimensional wave equation
$$u_{tt} = c^2(u_{xx} + u_{yy}), \ 0 < x < a, \ 0 < y < b, \ t > 0,$$

and the boundary conditions
$$u(0, y, t) = u(a, y, t) = 0,$$
$$u(x, 0, t) = u(x, b, t) = 0,$$

where $\lambda_{nm}^2 = (\frac{n\pi}{a})^2 + (\frac{m\pi}{b})^2$ and $a_{nm}$ and $b_{nm}$ are constants that can be determined if we prescribe the initial conditions $u(x,y,0)$ and $u_t(x,y,0)$.

4. Verify that

$$u(x,y,t) = \sum_{n=1}^{\infty} \sum_{m=1}^{\infty} a_{nm} e^{-k\lambda_{nm}^2 t} \sin\left(\frac{n\pi x}{a}\right) \sin\left(\frac{m\pi y}{b}\right),$$

satisfies the two-dimensional heat equation

$$u_t = k(u_{xx} + u_{yy}), \ 0 < x < a, \ 0 < y < b, \ \text{and } t > 0,$$

and the boundary conditions

$$\begin{aligned}
u(0,y,t) &= u(a,y,t) = 0, \\
u(x,0,t) &= u(x,b,t) = 0,
\end{aligned}$$

where $\lambda_{nm}^2 = (\frac{n\pi}{a})^2 + (\frac{m\pi}{b})^2$. The constants $a_{nm}$ can be determined if we prescribe the initial condition $u(x,y,0)$.

**Note**: It is assumed in the last two exercises that the series on the right-hand side and the series obtained by differentiating them term-by-term twice are uniformly convergent.

## 2.8 Families of Equipotential Surfaces

Let

$$f(x,y,z) = c, \tag{2.8.1}$$

be a one-parameter family of surfaces. We say that this family of surfaces is equipotential if there exists a potential function $\psi$ (a solution of Laplace's equation), such that $\psi$ is constant whenever $f$ is constant. There must therefore be a functional relation of the type

$$\psi(x,y,z) = F\{f(x,y,z)\}. \tag{2.8.2}$$

On differentiating (2.8.2) twice with respect to $x$, we get

$$\frac{\partial^2 \psi}{\partial x^2} = \frac{d^2 F}{df^2} \left(\frac{\partial f}{\partial x}\right)^2 + \frac{dF}{df} \frac{\partial^2 f}{\partial x^2}.$$

Similarly, on differentiating (2.8.2) twice with respect to $y$ and $z$ and adding them, we get

$$\nabla^2 \psi = F''(f)(\nabla f)^2 + F' \nabla^2 f.$$

Since $\nabla^2 \psi = 0$, the required necessary condition is

$$\frac{\nabla^2 f}{(\nabla f)^2} = -\frac{F''(f)}{F'(f)}. \qquad (2.8.3)$$

Therefore the condition that the surfaces (2.8.1) form a family of equipotential surfaces is that the quantity $\dfrac{\nabla^2 f}{|\nabla f|^2}$ is a function of $f$ alone denoted by $\chi(f)$, say. Therefore we have

$$\frac{d^2 F}{df^2} + \chi(f)\frac{dF}{df} = 0.$$

On integrating the previous ordinary differential equation, we get

$$\psi = F(f) = A \int e^{-\int \chi(f)df} df + B, \qquad (2.8.4)$$

where $A$ and $B$ are constants.

**Note**: The previously derived results including (2.8.4) are true even in two dimensions.

**Example 2.8.1**: Consider the family of surfaces

$$f(x, y, z) = x^2 + y^2 + z^2 = c, \quad c > 0,$$
$$\nabla^2 f = 6,$$
$$\nabla f = 2x\hat{i} + 2y\hat{j} + 2z\hat{k},$$
$$|\nabla f|^2 = 4f.$$

Therefore

$$\frac{\nabla^2 f}{|\nabla f|^2} = \frac{3}{2f} = \chi(f),$$
$$\psi = A \int e^{-\int \chi(f)df} df + B,$$
$$= -\frac{2A}{r} + B. \qquad \square$$

**Exercise 2.8.1**: Show that the surfaces

$$x^2 + y^2 + z^2 = cx^{2/3}$$

can form an equipotential family of surfaces, and find the general form of the potential function.

**Exercise 2.8.2**:  Show that $\psi = (x^2 + y^2 + z^2)^{-1/2}$ satisfies the three-dimensional Laplace's equation in any domain that does not contain the origin.

**Note**: In fact, this function is the fundamental solution of Laplace's equation in three dimensions. Observe that unlike in two dimensions, the fundamental solution in three dimensions tends to zero at infinity.

## 2.9   Kelvin's Inversion Theorem

**Theorem 2.9.1**:

If $\Phi = \Phi(r, \theta, \varphi)$ is a harmonic function, where $(r, \theta, \varphi)$ are the spherical polar coordinates, then $\tilde{\Phi} = \dfrac{a^2}{r} \Phi(\dfrac{a^2}{r}, \theta, \varphi)$ is also a harmonic function, where $a$ is a constant.

**Proof**: Let $R = a^2/r$. Then $\tilde{\Phi} = R\Phi(R, \theta, \varphi)$ .

Since $\Phi$ is harmonic, it satisfies the equation

$$\frac{\partial}{\partial r}\left(r^2 \frac{\partial \Phi}{\partial r}\right) + \frac{1}{\sin\theta}\frac{\partial}{\partial \theta}\left(\sin\theta \frac{\partial \Phi}{\partial \theta}\right) + \frac{1}{\sin^2\theta}\frac{\partial^2 \Phi}{\partial \varphi^2} = 0, \qquad (2.9.1)$$

which is the Laplace's equation in spherical polar coordinates, but for the factor $\dfrac{1}{r^2}$.

So $\Phi(R, \theta, \varphi)$ satisfies the same equation with $R$ replacing $r$

$$\frac{\partial}{\partial R}\left(R^2 \frac{\partial \Phi}{\partial R}\right) + \frac{1}{\sin\theta}\frac{\partial}{\partial \theta}\left(\sin\theta \frac{\partial \Phi}{\partial \theta}\right) + \frac{1}{\sin^2\theta}\frac{\partial^2 \Phi}{\partial \varphi^2} = 0. \qquad (2.9.2)$$

Now

$$r^2 \frac{\partial \tilde{\Phi}}{\partial r} = -a^2 \Phi(R, \theta, \varphi) - \frac{a^4}{r}\frac{\partial \Phi}{\partial R}(R, \theta, \varphi), \qquad (2.9.3)$$

$$\frac{\partial}{\partial r}\left(r^2 \frac{\partial \tilde{\Phi}}{\partial r}\right) = \frac{2a^4}{r^2}\frac{\partial \Phi}{\partial R} + \frac{a^6}{r^3}\frac{\partial^2 \Phi}{\partial R^2} = R\frac{\partial}{\partial R}\left(R^2 \frac{\partial \Phi}{\partial R}\right). \qquad (2.9.4)$$

Therefore

$$
\frac{\partial}{\partial r}\left(r^2\frac{\partial\tilde{\Phi}}{\partial r}\right) + \frac{1}{\sin\theta}\frac{\partial}{\partial\theta}\left(\sin\theta\frac{\partial\tilde{\Phi}}{\partial\theta}\right) + \frac{1}{\sin^2\theta}\frac{\partial^2\tilde{\Phi}}{\partial\varphi^2}
$$
$$
= R\left(\frac{\partial}{\partial R}\left(R^2\frac{\partial\Phi}{\partial R}\right) + \frac{1}{\sin\theta}\frac{\partial}{\partial\theta}\left(\sin\theta\frac{\partial\Phi}{\partial\theta}\right)\right.
$$
$$
\left. + \frac{1}{\sin^2\theta}\frac{\partial^2\Phi}{\partial\varphi^2}\right),
$$
$$
= 0, \tag{2.9.5}
$$

from (2.9.2).
Hence $\tilde{\Phi}(r,\theta,\varphi)$ is a harmonic function. □
**Corollary**: Let $\Phi_0(r,\theta,\varphi)$ be a harmonic function. Then

$$
\Phi(r,\theta,\varphi) = \Phi_0(r,\theta,\varphi) - \frac{a}{r}\Phi_0(\frac{a^2}{r},\theta,\varphi),
$$

is also harmonic such that $\Phi = 0$ on $r = a$. □
**Note**: The previous corollary has been used extensively to prove some important results that arise in the problems of flow past spherical boundaries in fluid dynamics.
**Exercise 2.9.1**: Show using the method of separation of variables that the general solution $\Phi$ of Laplace's equation in $(r,\theta,\varphi)$ coordinates is

$$
\Phi(r,\theta,\varphi) = \sum_{n=0}^{\infty}\left(\alpha_n r^n + \beta_n\frac{1}{r^{n+1}}\right)S_n(\theta,\varphi),
$$

where $S_n(\theta,\varphi) = \sum_{m=0}^{n}P_n^m(\zeta)(A_{nm}\cos m\varphi + B_{nm}\sin m\varphi)$, $\zeta = \cos\theta$, $P_n^m(\zeta)$ is the associated Legendre function, and $\alpha_n,\beta_n,A_{nm}$, and $B_{nm}$ are constants. Verify Kelvin's inversion theorem for the previous solution.
**Exercise 2.9.2**: A solution of Equation (2.9.1) that depends only on $r$ is called a radial solution of the Laplace's equation. Find such a solution.
**Note**: Harnack's theorem is also true for three dimensions.

# Appendix A

# Fourier Transforms and Integrals

## A.1 Fourier Integral Theorem

**Theorem A1**: Let $f(x)$ be a continuous and absolutely integrable function in $(-\infty, \infty)$. Then $\mathcal{F}(f) = F(\alpha)$ called the Fourier transform of $f(x)$ is defined as

$$F(\alpha) = \mathcal{F}(f) = \frac{1}{\sqrt{2\pi}} \int_{-\infty}^{\infty} f(x)e^{i\alpha x}dx.$$

Then

$$f(x) = \mathcal{F}^{-1}(F) = \frac{1}{\sqrt{2\pi}} \int_{-\infty}^{\infty} F(\alpha)e^{-i\alpha x}d\alpha,$$

is called the inverse Fourier transform of $F(\alpha)$. $\qquad\square$

## A.2 Properties of the Fourier Transform

1. $\mathcal{F}(af + bg) = a\mathcal{F}(f) + b\mathcal{F}(g)$, where $a$ and $b$ are constants.

2. $\mathcal{F}(f(x - c)) = e^{i\alpha c}\mathcal{F}(f(x))$, where $c$ is a real constant.

3. $\mathcal{F}(f(cx)) = \frac{1}{|c|}F(\frac{\alpha}{c})$ for $c \neq 0$, where $c$ is a real constant.

4. If $f$ and $f'$ are absolutely integrable and $f'$ is continuous, then $\mathcal{F}(f'(x)) = -i\alpha\mathcal{F}(f(x))$ (provided that $f \to 0$ as $|x| \to \infty$).

5. If $f$ is such that its $n$th derivative is continuous and $f$ and its first $n$ derivatives are absolutely integrable then
$\mathcal{F}(f^{(n)}(x)) = (-i\alpha)^n \mathcal{F}(f(x))$ (provided that $f$ and its $(n-1)$ derivatives tend to zero as $\mid x \mid \to \infty$).

6. $\mathcal{F}(\int^x f(s)ds) = \dfrac{iF(\alpha)}{\alpha}$.

## A.3   Examples

**Example A.3.1**: Find the Fourier transform of

$$f(x) = e^{-|x|}.$$

**Solution**: We have

$$F(\alpha) = \frac{1}{\sqrt{2\pi}} \int_{-\infty}^{\infty} e^{-|x|} e^{i\alpha x} dx,$$

$$= \frac{1}{\sqrt{2\pi}} \left( \int_{-\infty}^{0} e^{x} e^{i\alpha x} dx + \int_{0}^{\infty} e^{-x} e^{i\alpha x} dx \right),$$

$$= \frac{1}{\sqrt{2\pi}} \left( \frac{1}{1+i\alpha} + \frac{1}{1-i\alpha} \right),$$

$$= \sqrt{\frac{2}{\pi}} \left( \frac{1}{1+\alpha^2} \right). \qquad \square$$

**Example A.3.2**: Find the Fourier transform of

$$f(x) = e^{-x^2}.$$

**Solution**: We have

$$F(\alpha) = \frac{1}{\sqrt{2\pi}} \int_{-\infty}^{\infty} e^{-x^2} e^{i\alpha x} dx,$$

$$= \frac{1}{\sqrt{2\pi}} \int_{-\infty}^{\infty} e^{-(x^2 - i\alpha x + (i\alpha/2)^2) - (\alpha^2/4)} dx,$$

$$= \frac{e^{-\alpha^2/4}}{\sqrt{2\pi}} \int_{-\infty}^{\infty} e^{-(x - i\alpha/2)^2} dx.$$

Put $\eta = x - i\alpha/2$. Then $d\eta = dx$. Therefore

$$F(\alpha) = \frac{e^{-\alpha^2/4}}{\sqrt{2\pi}} \int_{-\infty}^{\infty} e^{-\eta^2} d\eta,$$

$$= \frac{e^{-\alpha^2/4}}{\sqrt{2\pi}} \sqrt{\pi} \, ,$$

$$= \frac{1}{\sqrt{2}} e^{-\alpha^2/4} \, .$$

Therefore

$$\mathcal{F}^{-1}(\frac{1}{\sqrt{2}} e^{-\alpha^2/4}) = e^{-x^2}. \qquad \square$$

**Exercise A.3.1**: Show that

$$\mathcal{F}^{-1}(e^{-|\alpha|y}) = \sqrt{\frac{2}{\pi}} \left( \frac{y}{y^2 + x^2} \right).$$

**Exercise A.3.2**: Show that

$$\mathcal{F}^{-1}(e^{-a\alpha^2}) = \frac{1}{\sqrt{2a}} \exp(-\frac{x^2}{4a}).$$

# A.4   Convolution Theorem

**Theorem A2**: Let $f(x)$ and $g(x)$ possess Fourier transforms. Define

$$(f * g)(x) = \frac{1}{\sqrt{2\pi}} \int_{-\infty}^{\infty} f(x - \xi)g(\xi)d\xi,$$

called the convolution of the functions $f$ and $g$ over the interval $(-\infty, \infty)$. Then

$$\mathcal{F}(f * g) = \mathcal{F}(f)\mathcal{F}(g).$$

## A.5    Vibrations of an Infinite String

Let us consider the problem of vibrations of an infinite string governed by the following equations

$$u_{tt} - c^2 u_{xx} = 0, \quad -\infty < x < \infty, \quad t > 0. \tag{A1}$$

The initial conditions are

$$u(x, 0) = f(x), \quad -\infty < x < \infty, \tag{A2}$$

and
$$u_t(x, 0) = g(x), \quad -\infty < x < \infty. \tag{A3}$$

Taking the Fourier transform of (A1) – (A3), we get

$$\frac{\partial^2 U}{\partial t^2}(\alpha, t) + (c\alpha)^2 U(\alpha, t) = 0, \quad t > 0, \tag{A4}$$

$$U(\alpha, 0) = F(\alpha), \tag{A5}$$

$$\frac{\partial U}{\partial t}(\alpha, 0) = G(\alpha), \tag{A6}$$

where $U, F,$ and $G$ are the Fourier transforms of $u, f,$ and $g$ respectively. The solution of (A4) subject to the conditions (A5) and (A6) is

$$U(\alpha, t) = \frac{1}{2}[F(\alpha) + \frac{1}{i\alpha c}G(\alpha)]e^{i\alpha ct} + \frac{1}{2}[F(\alpha) - \frac{1}{i\alpha c}G(\alpha)]e^{-i\alpha ct}. \tag{A7}$$

Therefore, taking the inverse Fourier transform of (A7), we get

$$
\begin{aligned}
u(x,t) &= \frac{1}{\sqrt{2\pi}} \int_{-\infty}^{\infty} e^{-i\alpha x} U(\alpha, t) d\alpha, \\
&= \frac{1}{2} \left( \frac{1}{\sqrt{2\pi}} \int_{-\infty}^{\infty} e^{-i\alpha(x-ct)} F(\alpha) d\alpha \right. \\
&\quad + \frac{1}{\sqrt{2\pi}} \int_{-\infty}^{\infty} e^{-i\alpha(x+ct)} F(\alpha) d\alpha \\
&\quad + \frac{1}{c\sqrt{2\pi}} \int_{-\infty}^{\infty} e^{-i\alpha(x-ct)} \frac{G(\alpha)}{i\alpha} d\alpha \\
&\quad \left. - \frac{1}{c\sqrt{2\pi}} \int_{-\infty}^{\infty} e^{-i\alpha(x+ct)} \frac{G(\alpha)}{i\alpha} d\alpha \right), \\
&= \frac{1}{2} [f(x - ct) + f(x + ct)] \\
&\quad - \frac{1}{2c} \int^{x-ct} g(s) ds + \frac{1}{2c} \int^{x+ct} g(s) ds, \\
&= \frac{1}{2} [f(x - ct) + f(x + ct)] + \frac{1}{2c} \int_{x-ct}^{x+ct} g(s) ds. \qquad (A8)
\end{aligned}
$$

This is nothing but d'Alembert's solution of wave equation that we derived earlier (refer to Equation (2.3.2)).

# Appendix B

# Additional Problems

1. Find the general solution of the following quasi-linear partial differential equations

   (i) $(x^2 + 3y^2 + 3z^2)p - 2xyq + 2xz = 0.$

   (ii) $zp + (z^2 - x^2)q = -x.$

   (iii) $(y + xz)p - (x + yz)q = x^2 - y^2.$

   (iv) $(x^2 - yz)p + (y^2 - xz)q = z^2 - xy.$

   (v) $x(y^2 + z)p - y(x^2 + z)q = (x^2 - y^2)z.$

2. Find the general solution of the following partial differential equations (refer to Theorem 1.4.2)

   (a) $(z - y)^2 u_x + zu_y + yu_z = 0.$

   (b) $x(y - z)u_x + y(z - x)u_y + z(x - y)u_z = 0.$

3. Consider the p.d.e. $yz_x - xz_y = 0.$ Show that this equation has a unique solution for the initial data curve $x_0(s) = s$, $y_0(s) = 0$, $z_0(s) = s^2$ ; has no solution for the initial data curve $x_0(s) = \cos s$, $y_0(s) = \sin s$, $z_0(s) = \sin s$; and has many solutions for $x_0(s) = \cos s$, $y_0(s) = \sin s$, $z_0(s) = 1$. Explain these cases in view of Theorem 1.10.1.

4. Show that $z = xy$ is a common solution of $f = pq - z = 0$ and $g = pq - xy = 0$, but $f$ and $g$ are not compatible. (Note: It is not enough for two partial differential equations to be compatible that they have just one solution in common.)

5. Solve the following Cauchy problems

    (i) $zz_x + yz_y = x$, $\quad x_0(s) = s$, $y_0(s) = 1$, $z_0(s) = 2s$.

    (ii) $p^2 + yq - z = 0$, $\quad x_0(s) = s$, $y_0(s) = 1$, $z_0(s) = \dfrac{s^2}{4} + 1$.

    (iii) $z = p^2 + q^2$, $\quad x_0(s) = s$, $y_0(s) = 0$, $z_0(s) = as^2$.
       For what values of $a$ is there a solution? Is it unique? Find all solutions.

    (iv) $pq = z$, $\quad x_0(s) = s$, $y_0(s) = \dfrac{1}{s}$, $z_0(s) = 1$. Show that there are two
       solutions.

6. Reduce the following partial differential equations into a canonical form and solve them

    (i) $xu_{xx} + 2x^2 u_{xy} = u_x - 1$.
       **Ans.:** $u(x, y) = x + F(x^2 - y) + G(y)$.

    (ii) $x^2 u_{xx} + 2x u_{xy} + u_{yy} = u_y$.
       **Ans.:** $u(x, y) = F(xe^{-y}) + G(xe^{-y})e^y$.

    (iii) $u_{xx} - 2u_{xy}\sin x - u_{yy}\cos^2 x - u_y \cos x = 0$.
       **Ans.:** $u(x, y) = F(y - x - \cos x) + G(y + x - \cos x)$.

    (iv) $u_{xx} + \dfrac{10}{3}u_{xy} + u_{yy} + \sin(x + y) = 0$.

       **Ans.:** $u(x, y) = F(x - 3y) + G(y - 3x) + \dfrac{3}{16}\sin(x + y)$.

    (v) $2x^2 u_{xx} + 5xy u_{xy} + 2y^2 u_{yy} + 8xu_x + 5yu_y = 0$.
       **Ans.:** $u(x, y) = \dfrac{x}{y^2}F\left(\dfrac{x^2}{y}\right) + G\left(\dfrac{y^2}{x}\right)$.

    Note: $F$ and $G$ occurring in the previous solutions are arbitrary $C^2$ functions.

7. Solve
$$\begin{aligned} u_{tt} - u_{xx} &= 0, \quad 0 < x < \infty, \; t > 0, \\ u(x, 0) &= \sin^3 x, \quad u_t(x, 0) = 0, \quad 0 \le x < \infty, \\ u(0, t) &= 0, \quad t > 0. \end{aligned}$$

8. Consider
$$\begin{aligned} u_{tt} - u_{xx} &= 0, \quad 0 < x < l, \quad t > 0, \\ u(x, 0) &= 0, \quad 0 \le x \le l, \\ u_t(x, 0) &= 0, \quad 0 \le x \le l, \\ u(0, t) &= f(t), \quad u(l, t) = g(t), \quad t > 0. \end{aligned}$$

Show that $v(x,t) = u(x,t) - \left(\frac{x}{l}\right) g(t) - \left(1 - \left(\frac{x}{l}\right)\right) f(t)$ reduces the previous problem to the case given in Exercise 2.6.1.

9. Solve

$$
\begin{aligned}
u_{tt} &= u_{xx}, \quad 0 < x < 1, \quad t > 0, \\
u(x,0) &= x^2(1-x)^2, \quad 0 \le x \le 1, \\
u_t(x,0) &= 0, \quad 0 \le x \le 1, \\
u_x(0,t) &= 0, \quad u_x(1,t) = 0, \quad t > 0.
\end{aligned}
$$

10. Find the solution of the problem

$$
u_{tt} = 4u_{xx}, \quad -\infty < x < \infty, \quad t > 0,
$$

$$
u(x,0) = \begin{cases} x/2, & 0 \le x \le 1/2, \\ (1-x)/2, & 1/2 \le x \le 1, \\ 0, & \text{elsewhere}, \end{cases}
$$

$$
u_t(x,0) = 0.
$$

Evaluate $u(1/2, 1/8)$ and $u(1,2)$.

11. Consider the p.d.e. $u_{tt} - u_{xx} + u = 0$. Show that

$$
u_1(x,t) = a\cos(x - \sqrt{2}t) \quad \text{and} \quad u_2(x,t) = a\cos(2x - \sqrt{5}t),
$$

where $a$ is a constant, are both its solutions. (However the two wave forms travel with different speeds of propagation $\sqrt{2}$ and $\sqrt{5}/2$, i.e., the waves with different wave lengths travel with different speeds, unlike in the case of Equation (2.3.1) where the speed of propagation is $\pm c$. This phenomenon is called *dispersion*.)

12. Solve by the method of separation of variables

$$
u_{xx} + u_{yy} = 0, \quad 0 < x < \pi, \; 0 < y < \pi,
$$

$$
u_x(0,y) = 0 = u_x(\pi,y), \quad 0 \le y \le \pi,
$$

$$
u(x,0) = 0, \; u(x,\pi) = g(x), \quad 0 \le x \le \pi.
$$

13. Derive the necessary condition for the Neumann problem

$$u_{xx} + u_{yy} = -F(x, y) \text{ in } D,$$
$$\frac{\partial u}{\partial n} = f(s) \text{ on } B,$$

where $B$ is the boundary of the domain $D$.

14. Find the solution of the initial-boundary value problem

$$u_t - u_{xx} + u = 0, \quad 0 < x < \pi, \quad t > 0$$
$$u(0, t) = u_x(\pi, t) = 0, \quad t \geq 0$$
$$u(x, 0) = x(\pi - x), \quad 0 < x < \pi.$$

(Hint: Substitute $v(x, t) = e^t u(x, t)$.)

15. Assuming convergence of the following series so that term-by-term differenti-ation once with respect to $t$ and twice with respect to $x$ are allowed, show that

$$u(x, t) = \sum_{k=0}^{\infty} \frac{1}{(2k)!} x^{2k} \frac{d^k}{dt^k} (e^{-1/t^2}),$$

satisfies $u_t = u_{xx}$ for $-\infty < x < \infty$, $t > 0$ and $u(x, 0) = 0$ for $-\infty < x < \infty$. Note: The solution $u(x, t)$ given in this problem is unbounded. Hence, the solution of Equations (2.5.1) and (2.5.2) is not unique unless we assume $u(x, t)$ is bounded.

16. Solve

$$u_t = u_{xx}, \quad 0 < x < 1, \ t > 0,$$
$$u(x, 0) = x^2(3 - 2x), \quad 0 \leq x \leq 1,$$
$$u_x(0, t) = 0, \quad u_x(1, t) = 0, \quad t > 0.$$

17. Solve

$$u_t - u_{xx} = e^{-t} \sin^2 x, \quad 0 < x < \pi, \ t > 0,$$
$$u(x, 0) = 0, \quad 0 \leq x \leq \pi,$$
$$u_x(0, t) = 0 = u_x(\pi, t), \quad t > 0.$$

18. In the heat conduction problem given in Equations (2.5.1) and (2.5.2), show that if $f$ is bounded and continuous in $\mathbb{R}$, then the solution given in Equations (2.5.3) is bounded and infinitely differentiable.

19. If $u(x, t)$ satisfies the heat conduction equation then show that $v(x, y, t) = u(x, t)u(y, t)$ satisfies the equation $v_t = v_{xx} + v_{yy}$, i.e., the heat conduction equation in two dimensions. Check whether such a result is true in the case of the wave equation.

20. Determine the regions where the following second order semi-linear partial differential equation in three independent variables $x$, $y$, and $z$ is of hyperbolic, elliptic, or parabolic type

$$u_{xx} - 2x^2 u_{xz} + u_{yy} + u_{zz} = 0.$$

# Suggested Reading

1 **F. John** *Partial Differential Equations*, Springer Verlag (1975).

2 **Erich Zauderer** *Partial Differential Equations of Applied Mathematics*, 2nd edition, Wilcy-Interscience publication (1989).

3 **Lawrence C. Evans** *Partial differential equations*, Graduate Studies in Mathematics, 19, American Mathematical Society, Providence, RI (1998).

4 **Robert C. McOwen** *Partial Differential Equations: Methods and Applications*, Second Edition, Prentice Hall (2003).

5 **R. Courant and D. Hilbert** *Methods of Mathematical Physics*, Vol. 2, Wiley-Interscience (1962).

6 **H.F. Weinberger** *A First Course in Partial Differential Equations*, Blaisdell, New York (1965).

7 **G.F.D. Duff and D. Naylor** *Differential Equations of Applied Mathematics*, Wiley, New York (1966).

8 **I.N. Sneddon** *Elements of Partial Differential Equations*, McGraw-Hill, New York (1957).

9 **P. Garabedian** *Partial Differential Equations*, 2nd edition Chelsea, New York (1986).

10 **S.L. Sobolev** *Partial Differential Equations of Mathematical Physics*, Addison-Wesley, Reading, Mass (1964).

11 **W.F. Ames** *Non-linear Partial Differential Equations in Engineering*, Academic Press (1965).

12 **Michael Spivak** *A Comprehensive Introduction to Differential Geometry*, Vol. 5, 2nd edition, Publish or Perish Inc., Wilmington, Delaware (1979).

13 **Miroslaw Krżyzański** *Partial Differential Equations of Second Order*, Vol. I & II, Monografie Matematyczne, PWN-Polish Scientific Publishers (1971).

14 **Tyn Myint-U** *Partial Differential Equations of Mathematical Physics*, Elsevier (1973).

# Index